Using the R Commander

A Point-and-Click
Interface for R

Chapman & Hall/CRC
The R Series

Series Editors

John M. Chambers
Department of Statistics
Stanford University
Stanford, California, USA

Torsten Hothorn
Division of Biostatistics
University of Zurich
Switzerland

Duncan Temple Lang
Department of Statistics
University of California, Davis
Davis, California, USA

Hadley Wickham
RStudio
Boston, Massachusetts, USA

Aims and Scope

This book series reflects the recent rapid growth in the development and application of R, the programming language and software environment for statistical computing and graphics. R is now widely used in academic research, education, and industry. It is constantly growing, with new versions of the core software released regularly and more than 7,000 packages available. It is difficult for the documentation to keep pace with the expansion of the software, and this vital book series provides a forum for the publication of books covering many aspects of the development and application of R.

The scope of the series is wide, covering three main threads:
- Applications of R to specific disciplines such as biology, epidemiology, genetics, engineering, finance, and the social sciences.
- Using R for the study of topics of statistical methodology, such as linear and mixed modeling, time series, Bayesian methods, and missing data.
- The development of R, including programming, building packages, and graphics.

The books will appeal to programmers and developers of R software, as well as applied statisticians and data analysts in many fields. The books will feature detailed worked examples and R code fully integrated into the text, ensuring their usefulness to researchers, practitioners and students.

Published Titles

Spatial Microsimulation with R, *Robin Lovelace and Morgane Dumont*

Statistics in Toxicology Using R, *Ludwig A. Hothorn*

Stated Preference Methods Using R, *Hideo Aizaki, Tomoaki Nakatani, and Kazuo Sato*

Using R for Numerical Analysis in Science and Engineering, *Victor A. Bloomfield*

Event History Analysis with R, *Göran Broström*

Computational Actuarial Science with R, *Arthur Charpentier*

Statistical Computing in C++ and R, *Randall L. Eubank and Ana Kupresanin*

Basics of Matrix Algebra for Statistics with R, *Nick Fieller*

Reproducible Research with R and RStudio, Second Edition, *Christopher Gandrud*

R and MATLAB®, *David E. Hiebeler*

Nonparametric Statistical Methods Using R, *John Kloke and Joseph McKean*

Displaying Time Series, Spatial, and Space-Time Data with R, *Oscar Perpiñán Lamigueiro*

Programming Graphical User Interfaces with R, *Michael F. Lawrence and John Verzani*

Analyzing Sensory Data with R, *Sébastien Lê and Theirry Worch*

Parallel Computing for Data Science: With Examples in R, C++ and CUDA, *Norman Matloff*

Analyzing Baseball Data with R, *Max Marchi and Jim Albert*

Growth Curve Analysis and Visualization Using R, *Daniel Mirman*

R Graphics, Second Edition, *Paul Murrell*

Introductory Fisheries Analyses with R, *Derek H. Ogle*

Data Science in R: A Case Studies Approach to Computational Reasoning and Problem Solving, *Deborah Nolan and Duncan Temple Lang*

Multiple Factor Analysis by Example Using R, *Jérôme Pagès*

Using the R Commander
A Point-and-Click Interface for R

John Fox
McMaster University
Hamilton, Ontario, Canada

CRC Press
Taylor & Francis Group
Boca Raton   London   New York

CRC Press is an imprint of the
Taylor & Francis Group, an **informa** business

A CHAPMAN & HALL BOOK

CRC Press
Taylor & Francis Group
6000 Broken Sound Parkway NW, Suite 300
Boca Raton, FL 33487-2742

© 2017 by Taylor & Francis Group, LLC
CRC Press is an imprint of Taylor & Francis Group, an Informa business

No claim to original U.S. Government works

ISBN 13: 978-1-4987-4190-3 (pbk)

Visit the Taylor & Francis Web site at
http://www.taylorandfrancis.com

and the CRC Press Web site at
http://www.crcpress.com

To the memory of my mentor and friend,
Mel Guyer.

Contents

Preface

The R Commander is a point-and-click *graphical user interface (GUI)* for R, providing access to R statistical software through familiar menus and dialog boxes instead of by typing potentially arcane commands. I expect that this book, which explains how to use the R Commander, will be of interest to students and instructors in introductory and intermediate-level statistics courses, to researchers who want to use R without having to contend with writing commands, and to R users who will eventually transition to the command-line interface but who prefer to start more simply.

In particular, in a basic statistics course, the central goal (in my opinion) should be to teach fundamental statistical ideas—distribution, statistical relationship, estimation, sampling variation, observational vs. experimental data, randomization, and so on. One doesn't want a basic statistics course to devolve into an exercise in learning how to write commands for statistical software, letting the software tail wag the statistical dog. I initially wrote the R Commander for this reason: to provide a transparent, intuitive, point-and-click graphical user interface to R, implemented using familiar menus and dialog boxes, running on all commonly used operating systems (Windows, Mac OS X, and Linux/Unix systems), and distributed and installed as a standard R package—called the **Rcmdr** package.

Although it was originally intended for use in basic statistics classes, the current capabilities of the R Commander, described in the various chapters of this book, extend well beyond basic statistics. The current version of the **Rcmdr** package contains nearly 15,000 lines of R code (exclusive of comments, blank lines, documentation, etc.). The R Commander is, moreover (like R itself), designed to be extensible by *plug-in packages*, standard R packages that augment or modify the R Commander's menus and dialogs (see Chapter 9).

I caught wind of R in the late 1990s and incorporated it shortly thereafter in my teaching—in a graduate social statistics course in applied regression analysis and generalized linear models. In 2002, I published a book on using R (and S-PLUS) for applied regression analysis (Fox, 2002).[1]

I also wanted to use R to teach basic statistics to social science graduate students and undergrads, but I felt that R's command line interface was an obstacle. I expected that eventually someone would introduce a graphical user interface to R, but none materialized, and so, around 2002, I resolved to tackle this task myself. After some experimentation, I decided to use the Tcl/Tk GUI builder because the basic R distribution comes with the **tcltk** package, which provides an R interface to Tcl/Tk. This choice permitted me to write an R GUI—the R Commander—that runs on all operating systems supported by R and that is coded entirely in R, minimizing the necessity to install additional software.

This book provides background information on R and the R Commander. I explain how you can obtain and install R and the R Commander on your computer. Finally, I show you how to use the R Commander to perform a variety of common statistical tasks.

I hope that the book proves useful to you, and I invite you to contact me at `jfox@mcmaster.ca` with corrections, comments, and suggestions.

---

[1]The second edition of this book (Fox and Weisberg, 2011) focuses solely on R and is coauthored with Sanford Weisberg.

Acknowledgments

Michael Friendly, Allison Leanage, and several anonymous reviewers contributed helpful suggestions on a draft of this book.

I'm grateful to the numerous individuals whose contributions to the **Rcmdr** package are acknowledged in the package documentation (see the *Help > About Rcmdr* menu). In particular, Richard Heiberger made a number of contributions to the early development of the R Commander, not least of which was the original code for rendering R Commander plug-in packages self-starting. Miroslav Ristic substantially improved the code for the probability distribution dialogs. Milan Bouchet-Valat joined me as a developer of the **Rcmdr** package in 2013 and helped to modernize the R Commander interface and to adapt it better to various computing platforms.

Peter Dalgaard, a member of the R Core team, made a critical indirect contribution to the R Commander—in effect, making it possible—by incorporating the **tcltk** package in the standard R distribution. Similarly, Milan and I used Philippe Grosjean's **tcltk2** package to enhance the R Commander interface, and Philippe has been a valuable source of information on Tcl/Tk. Brian Ripley, another member of R Core, has been very generous with his time, helping me to solve a variety of problems in the development of the R Commander.

I'd also like to thank John Kimmel, my editor at Chapman & Hall/CRC Press, Shashi Kumar (for his LaTeX expertise), and the production staff at Chapman & Hall/CRC Press, for all their help and encouragement.

Finally, my work on the R Commander was partly supported by grants from the Social Sciences and Humanities Research Council of Canada, and from McMaster University through the Senator William McMaster Chair in Social Statistics.

1

Introducing R and the R Commander

This chapter introduces R and the R Commander, explaining what they are and where they came from. The chapter also outlines the contents of the book, shows how to access the web site for the book, and describes the typographical conventions used in the text.

1.1 What Are **R** and the **R Commander**?

The R Commander—the subject of this book—is a point-and-click *graphical user interface* (*GUI*) for R, allowing you to use R statistical software through familiar menus and dialog boxes instead of by typing commands. Throughout the book, I assume that the statistical methods covered are familiar to you—or that you're concurrently learning them in a statistics class or by independently reading a complementary statistics text. The object of the book is to show you how to perform data analysis with the R Commander employing common statistical methods, not to teach the statistical methods themselves.

An implication of this approach is that you should feel free to skip those parts of the book that take up statistical methods with which you're unfamiliar. For example, most of the material in Chapter 7, on working with statistical models in the R Commander, is beyond the level of a typical basic statistics course. Sections that deal with relatively advanced or difficult material are marked with asterisks.

R is highly capable, free, open-source statistical software. Although it is hard to know with any certainty how many people use R, it is—for example, judging by Internet traffic— possibly the most popular statistical software in the world. R is, in any event, very widely used, and its use is growing rapidly!

R incorporates a programming language that is finely adapted to the development of statistical applications. R descends from the S programming language, originally developed in the 1980s by statisticians and computer scientists at Bell Labs, led by John Chambers (see, e.g., Becker et al., 1988). Indeed, R can be regarded as a dialect of S. Eventually incorporated in a commercial product called S-PLUS, S was popular among statisticians prior to the development of R. At present, the free, open-source R has entirely eclipsed its commercial cousin S-PLUS.

R is *free* software in Richard Stallman's famously dual sense of the term (Stallman, 2002):[1] It is free in the obvious sense of being costless, but also in the deeper sense that users may freely modify and distribute R. Moreover, R is licensed under the Free Software Foundation's General Public License (GPL)—a "viral copy-left" that prevents individuals or companies from restricting users' freedom to modify further and redistribute R. Freedom in the second sense essentially presupposes that R is *open source*: that is, that R is distributed not only as an executable program but also that the source code for R (written in a variety of programming languages, including in R itself) is available to interested users. For more

---

[1] Richard Stallman is the founder of the Free Software Foundation, with which the R Project for Statistical Computing is associated.

information about R, visit the R web site at https://www.r-project.org/. The R Commander is also free, open-source software distributed under the GPL.

Analyzing data with R doesn't necessarily entail writing programs in the R language, because the basic R distribution comes with impressive built-in statistical functionality. The capabilities of the standard R distribution, however, are greatly extended by (as I write this) nearly 8000 user-contributed R add-on *packages*, freely available on the Internet through the Comprehensive R Archive Network (abbreviated CRAN, and alternatively pronounced as "kran" or "see-ran"; see https://cran.r-project.org/). Moreover, roughly 1000 additional R packages are available through the closely associated Bioconductor Project (http://bioconductor.org/), which develops software primarily for bioinformatics (genomics).

Whether you write your own R programs or use pre-packaged programs, standard data analysis in R consists of typing commands in the R language. As a simple example, to compute the mean of the variable income, you might type the command `mean(income)`, invoking the standard R `mean` *function* (program). Similarly, to perform a linear least-squares regression of income on years of education and years of labor-force experience, you might issue the command `lm(income ~ education + experience)`, employing the `lm` (linear-model) function. Learning to write R commands like these is an important skill and ultimately is the most efficient way to use R (see Section 1.4), but it can present a formidable obstacle to new, occasional, or casual users of R.

1.2   A Brief History of R and the R Commander

R began around 1990 as the personal project of Robert Gentleman and Ross Ihaka, two statisticians then at Auckland University in New Zealand (see Ihaka and Gentleman, 1996). Gentleman and Ihaka in effect grafted the syntax of the pre-existing statistical programming language S onto the Scheme dialect of Lisp, a programming language usually associated with work in artificial intelligence. This turned out to be a propitious choice, because, as mentioned, the S language was already widely used by statisticians.

Eventually Ihaka and Gentleman advertised their work on the Internet, attracting several other developers to the project, including John Chambers, the principal developer of S. Then, in 1997, the *R Project for Statistical Computing* was formalized, with a Core team of nine developers, a number that has since expanded to 20, many of whom are significant figures in the field of statistical computing. The R Core team is responsible for the continued development and maintenance of the basic R distribution.

As I explained, R is distributed under the free-software General Public License. The copyright to R is held by the R Foundation, which comprises the members of the R Core team along with about a dozen other individuals; I'm an elected member of the R Foundation.

The growth of R has been nothing short of amazing. Figure 1.1 shows, for example, the expansion of the CRAN R package archive over the 14-year period for which I was able to obtain data.[2] The horizontal axes of the graph record R versions and corresponding dates, while the vertical axes show the number of CRAN packages on a logarithmic scale, so that a linear trend represents exponential growth.[3] The line on the graph was fit by least-squares

---

[2]This graph is updated from Fox (2009), where I discuss the social organization and trajectory of the R Project.

[3]If you're unfamiliar with logs, don't be concerned: The essential point is that the scale gets more compressed as the number of packages grows, so that, for example, the distance between 100 and 200 packages on the log scale is the *same* as the distance between 200 and 400, and the same as the distance between 400 and 800—all of these equal distances represent *doubling* the number of packages.

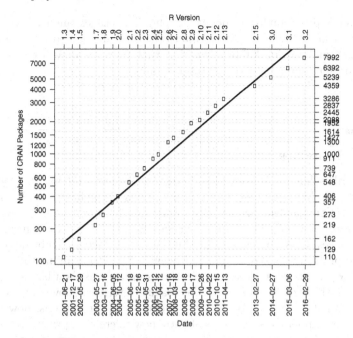

FIGURE 1.1: The growth of CRAN. The vertical axes show the number of CRAN packages on a log scale, while dates and corresponding minor R versions are shown at the bottom and top of the graph, respectively. The line on the graph was fit to the points by least-squares regression. Two R versions, 1.6 and 2.14, are omitted because their recorded dates were very close to the dates of the previous versions.
Source: Updated from Fox, "Aspects of the social organization and trajectory of the R Project," *The R Journal,* 1(2): 5–13, 2009.
Source of Data: Downloaded from https://svn.r-project.org/R/branches/ on 2016-03-03.

regression.[4] You can see from Figure 1.1 that, while the growth of CRAN was originally approximately exponential, its rate of growth has more recently slowed down. The slope of the least-squares line suggests that, on average over this period, CRAN expanded at a rate of about 35 percent a year.

As I mentioned in the Preface to this book, I began to work on the R Commander around 2002, and I contributed version 0.8-2 of the **Rcmdr** package to CRAN in May 2003. The first "non-beta" version, 1.0-0, appeared two years later, and was described in a paper in the *Journal of Statistical Software* (Fox, 2005), an on-line journal of the American Statistical Association. In March 2016, when I wrote the chapter you're reading, that paper had been downloaded nearly 140,000 times—despite the fact that it was more than 10 years out of date!

I have continued to develop the R Commander in the intervening period: Version 1.1-1, which also appeared in 2005, introduced the capability to translate the R Commander interface into other languages, a feature supported by R itself, and there are now 18 such translations (counting Chinese and simplified Chinese as separate translations). In 2007,

---

[4]Again, if you're unfamiliar with the method of least squares, don't worry: You'll almost surely study the topic in your basic statistics course. The essential idea is that the line comes as close (in a sense) to the points on average as possible.

Version 1.3-0 first made provision for R Commander plug-in packages, and there are currently about 40 such plug-ins on CRAN. In 2013, Milan Bouchet-Valat joined me as a developer of the R Commander, and version 2.0-0, released in that year, featured a revamped, more consistent interface—for example, featuring tabbed dialogs.

1.3 Chapter Synopses

Chapter 2 describes how to download R from the Internet and install it and the R Commander on Windows, Mac OS X, and Linux/Unix systems. If you have already successfully installed R and the R Commander, then feel free to skip the chapter. There is, however, some troubleshooting information, to which you can make reference if you experience a problem.

Chapter 3 introduces the R Commander graphical user interface (GUI) by demonstrating its use for a simple problem: constructing a contingency table to examine the relationship between two categorical variables. In developing the example, I explain how to start the R Commander, describe the structure of the R Commander interface, show how to read data into the R Commander, how to modify data to prepare them for analysis, how to draw a graph, how to compute numerical summaries of data, how to create a printed report of your work, how to edit and re-execute commands generated by the R Commander, and how to terminate your R and R Commander session—in short, the typical work flow of data analysis using the R Commander. I also explain how to customize the R Commander interface.

Chapter 4 shows how to get data into the R Commander from a variety of sources, including entering data directly at the keyboard, reading data from a plain-text file, accessing data stored in an R package, and importing data from an Excel or other spreadsheet, or from other statistical software. I also explain how to save and export R data sets from the R Commander, and how to modify data—for example, how to create new variables and how to subset the active data set.

Chapter 5 explains how to use the R Commander to compute simple numerical summaries of data, to construct and analyze contingency tables, and to draw common statistical graphs. Most of the statistical content of the chapter is covered in a typical basic statistics course, although a few topics, such as quantile-comparison plots and smoothing scatterplots, are somewhat more advanced.

Chapter 6 shows how to compute simple statistical hypothesis tests and confidence intervals for means, for proportions, and for variances, along with simple nonparametric tests, a test of normality, and correlation tests. Many of these tests are typically taken up in a basic statistics class, and, in particular, tests and confidence intervals for means and proportions are often employed to introduce statistical inference.

Chapter 7 explains how to fit linear and generalized linear regression models in the R Commander, and how to perform additional computations on regression models once they have been fit to data.

Chapter 8 explains how to use the R Commander to perform computations on probability distributions, to graph probability distributions, and to conduct simple random simulations.

Chapter 9 explains how to use R Commander *plug-in packages*. The capabilities of the R Commander are substantially augmented by the many plug-in packages for it that are available on CRAN. Plug-ins are R packages that add menus, menu items, and dialog boxes to the R Commander. I show you how to install plug-in packages, and illustrate the application of R Commander plug-ins by using the **RcmdrPlugin.TeachingDemos** package and the **RcmdrPlugin.survival** package as examples.

An appendix to the book displays the complete set of R Commander menus, along with cross-references to the text.

1.4 What's Next?

If you become a frequent user of R, you'll likely graduate from the R Commander to writing your own R commands and possibly your own R programs. There are several reasons to employ the command-line interface to R in preference to a GUI like the R Commander:

- The R Commander GUI provides access to only a small fraction of the capabilities of R and the many R packages available on CRAN. To take full advantage of R, therefore, you'll have to learn to write commands.

- Even if you limit yourself to the capabilities in the R Commander and its various plug-ins, frequent users of R find the command-line interface more efficient. Once you remember the various commands and their arguments, you'll learn to work more quickly at the command line than in a GUI.

- You'll find that a little bit of programming goes a long way. Writing simple scripts and programs is often the quickest and most straightforward way to perform data management tasks, for example.

If you do decide to learn to use R via the command-line interface, there are many books and other resources to help you. For example, I and Sanford Weisberg have written a text (Fox and Weisberg, 2011) that introduces R, including R programming, in the context of applied regression analysis. See the *Documentation* links on the R home page at https://www.r-project.org/ for many alternative sources, including free resources.

The R Commander is designed to facilitate the transition to command-line use of R: The commands produced by the R Commander are visible in the *R Script* tab. The contents of the *R Script* tab may be saved to a file and reused, either in the R Commander or in an R programming editor. Similarly, the dynamic document produced in the R Commander *R Markdown* tab may be saved, edited, and executed independently of the R Commander. These features are briefly discussed in Chapter 3.

Both the Windows and Mac OS X implementations of R come with simple programming editors, but I strongly recommend the RStudio *interactive development environment* (*IDE*) for command-line use of R. RStudio incorporates a powerful programming editor and is ideal both for routine data analysis in R and for R programming, including the development of R packages—and RStudio supports R Markdown documents. Like R and the R Commander, RStudio is free, open-source software: Visit the RStudio web site at https://www.rstudio.com/products/rstudio/ for details, including extensive documentation.

1.5   Web Site

I have created a web site to support this book with a variety of resources, including:

- all the data files used in examples that appear in the text

- detailed (and potentially updated) installation instructions, including trouble-shooting information, beyond the instructions in Chapter 2

- information about significant updates to the R Commander following the publication of this book, along with errata correcting errors in the book (as they, almost inevitably, reveal themselves)

- a manual for authors of R Commander plug-in packages

A note about software versions: Although some of the "screenshots" and output in this book were produced with earlier versions, the book is current as of version 3.2.3 of R and version 2.1-4 of the **Rcmdr** package. Significant changes to the R Commander or changes to R that affect the R Commander will be addressed on the web site for the book.

Chapman and Hall maintains a link to the web site for the book at https://www.crcpress. com/Using-the-R-Commander-A-Point-and-Click-Interface-for-R/Fox/p/book/9781498741903, which can also be accessed at http://tinyurl.com/RcmdrBook.

1.6   Typographical Conventions

Different typefaces and fonts are used to distinguish the following elements:

- Computer software, such as operating systems (Windows, Mac OS X, Linux) and statistical software (R, the R Commander) are shown in a sans serif typeface.

- Graphical user interface components, such as menus (the *Edit* menu) and windows (the *R Console* window, the *Two-Way Table* dialog box), are shown in an *italic typeface*.

- Submenu and menu item selection is indicated by > (a greater than sign). Thus, for example, *Statistics > Summaries > Numerical summaries...* means "left-click on the *Statistics* menu, then on the *Summaries* submenu, and finally on the *Numerical summaries...* menu item." Three dots following a menu item (as in *Numerical summaries...*) indicates that selecting the item leads to a dialog box, rather than performs a direct action. I will usually omit the three dots from menu items, however.

- Keys (e.g., *Tab*) and key combinations (*Ctrl-c*) are also shown in an *italic typeface*. The key combination *Ctrl-c*, for example, means "hold down the *Ctrl* key and press the *c* key."

- R packages (such as the **Rcmdr** and **car** packages) are shown in **boldface**.

- Text meant to be typed directly (such as the R command `library(Rcmdr)`, or text to be typed into an R Commander dialog box) is shown in a `typewriter font`, as is R output, and as are the names of R objects, such as functions (`mean`), data sets (`States`), and variables in data sets (`pay`). Generic text (e.g., *`variable-name`*) meant to be replaced with specific text (e.g., `income`) is shown in an *`italic typewriter font`*.

- Files (`GSS.csv`) and file paths (`C:\Program Files\R\R-`*`x.y.z`*`\`) are also shown in `typewriter font`, with generic text again in *`typewriter italics`*.

- Internet URLs (addresses) are shown in a sans serif typeface (e.g., https://cran.r-project.org).

- When important, possibly unfamiliar, terms are introduced, they are set in *italics* (e.g., *rectangular data set*).

2

Installing *R* and the *R Commander*

This chapter describes how to download R from the Internet and install it and the R Commander on Windows, Mac OS X, and Linux/Unix systems. If you have already successfully installed R and the R Commander, then feel free to skip the chapter. There is, however, some troubleshooting information, to which you can make reference if you experience a problem.

2.1 Acquiring and Installing **R** and the **R Commander**

More detailed (and potentially more up-to-date) information on installing R and the **Rcmdr** package appears on the web site for this book. Please consult the web site if the information provided here proves insufficient, or if you encounter difficulties not discussed here.

R and R packages, like the **Rcmdr** package, are available on the Internet from CRAN (the Comprehensive R Archive Network—see Chapter 1) at https://cran.r-project.org. It is best not to download R and R packages directly from the main CRAN web site, however, but rather to use a CRAN *mirror site*. A link to a list of CRAN mirrors appears at the upper left of the CRAN home page (the top of which is shown in Figure 2.1). I suggest that you use the first "0-Cloud" mirror, which is generally both reliable and fast.

Regardless of whether you are a Windows user, a Mac OS X user, or a Linux/Unix user, I recommend that you install the current version of R, say R version $x.y.z$. In this generic version number, "x" represents the *major version*, "y" the *minor version*, and "z" the *patch version* of R. A new minor version of R, $x.y.0$, is released by the R Core team each spring, and patch versions are released as needed, typically to fix bugs. Major versions appear infrequently, and only when substantial modifications are made to the base R software. As I'm writing this book, the current version of R is 3.2.3.[1]

There are (at most) five steps to installing R and the R Commander:

1. Download and install R.

2. On Mac OS X only, download and install the XQuartz windowing software.

3. Start R and install the **Rcmdr** package.

4. Load the **Rcmdr** package and, when asked, allow it to install additional packages.

5. If desired, optionally download and install Pandoc and LaTeX software for producing enhanced reports (as described in Section 3.6).

Specific instructions for Windows, Mac OS X, and Linux/Unix systems follow.

---

[1]The book, of course, was written over a period of time; R 3.2.3 was current when I was finalizing the text.

The Comprehensive R Archive Network
CRAN Mirrors What's new? Task Views Search *About R* R Homepage The R Journal	**Download and Install R** Precompiled binary distributions of the base system and contributed packages, **Windows and Mac** users most likely want one of these versions of R: • Download R for Linux • Download R for (Mac) OS X • Download R for Windows R is part of many Linux distributions, you should check with your Linux package management system in addition to the link above.

FIGURE 2.1: The top of the Comprehensive R Archive Network (CRAN) home page (on 2015-10-04).

2.2 Installing **R** and the **R Commander** under **Microsoft Windows**

From the home page of the mirror you selected, click on the link *Download R for Windows*, which appears near the top of the page. Then click on *install R for the first time*, and subsequently on *Download R* x.y.z *for Windows*.

Once it is downloaded, double-click on the R installer. On Windows 10, you will see a frightening message: "Windows protected your PC. Windows SmartScreen prevented an unrecognized app from starting. Running this app might put your PC at risk." Click on *More info* and then on the *Run anyway* button. (Why not live dangerously?) Windows issues this warning if your user-account controls are at the default settings and you download software that's not from the Windows App Store.

You may take all of the defaults in the R installer, but I suggest that you make the following modifications:

- Instead of installing R in the standard location, typically `C:\Program Files\R\R-x.y.z\` (for R version $x.y.z$), you may use `C:\R\R-x.y.z\`. This will allow you to install packages in the main R package library on your computer without running R with administrator privileges.

  If you install R into `C:\Program Files\R\R-x.y.z\`, and you run R without administrator privileges, you will see a message like the following:

  ```
  Warning in install.packages("Rcmdr") :
  'lib = "C:/Program Files/R/R-x.y.z/library"' is not writable
  ```

  and R will ask to install packages into a personal library under your Windows user account. That works too, but the installed packages will be available only to your account.

- On R for Windows, the R Commander works best with the "single-document interface" (or SDI). Under the default "multiple-document interface" (MDI), the main R Commander window and its various dialog boxes will not be contained in the master R window and may not stay on top of this window.[2]

In the *Startup options* screen of the R installer, select *Yes (customized startup)*. Then select the *SDI (single-document interface)* in preference to the default *MDI (multiple-document interface)*; feel free to make other changes, but you may take all the remaining defaults.

The key steps in the installation process, where I recommend that you make non-default choices, are shown in Figure 2.2.

Once it is installed, start R in the standard manner—for example, by double-clicking on its desktop icon, or by selecting it from the Windows start menu. If you are running R on a 64-bit Windows computer (and almost all current computers are 64-bit), *both* 64-bit and 32-bit versions of R will be installed. You can use either, but I suggest that you use the 64-bit version, and feel free to delete the 32-bit R icon from your desktop.[3]

The easiest way to install the **Rcmdr** package, if you have an active Internet connection, is via the *Packages > Install package(s)* menu in the *R Console* (that is, click on the *Packages* menu, select the menu item *Install package(s)*, and pick the **Rcmdr** package from the long alphabetized list of R packages available on CRAN), or via the command `install.packages("Rcmdr")` typed at the > command prompt in the *R Console* (followed by pressing the *Enter* key). In either case, R will ask you to select a CRAN mirror for the session; I suggest that you again pick the first "0-Cloud" mirror. R will install the **Rcmdr** package, along with a number of other R packages that the R Commander requires to get started.

When you first load the **Rcmdr** package with the command `library(Rcmdr)`, it will offer to download and install additional packages; allow it to do so.

2.2.1 Troubleshooting

Installing R and the R Commander on Windows systems is generally straightforward. Occasionally, and unpredictably, an R package required by the **Rcmdr** package fails to be installed—possibly because the package is missing from the CRAN mirror that you used—and the R Commander can't start. Under these circumstances, there is typically an informative error message about the missing package.

The simple solution is to install the missing package (or packages, if there are more than one) directly. For example, if the **car** package is missing, you can install it via the R command `install.packages("car")`, or from the *R Console Packages* menu, possibly selecting a different CRAN mirror from the one that you used initially.

Sometimes when users save the R workspace upon exiting from R, the R Commander will fail to work properly in a subsequent session. As I will explain in Section 3.8, I recommend *never* saving the R workspace to avoid these kinds of problems, and if you exit from R via the R Commander menus (*File > Exit > From Commander and R*), the R workspace will not be saved.

---

[2]To clarify, the R Commander works with *both* the SDI and the MDI, but it is more convenient to use it with the SDI.

[3]Occasionally, on Windows 8 systems, the 64-bit version of R seems incompatible with viewing HTML files for reports (as described in Section 3.6). In these cases, you can use the 32-bit version of R in preference to the 64-bit version. The principal advantage of the 64-bit version of R is that it permits the analysis of larger data sets, but it is unusual to use the R Commander to analyze bigger data sets than can be accommodated by the 32-bit version of R.

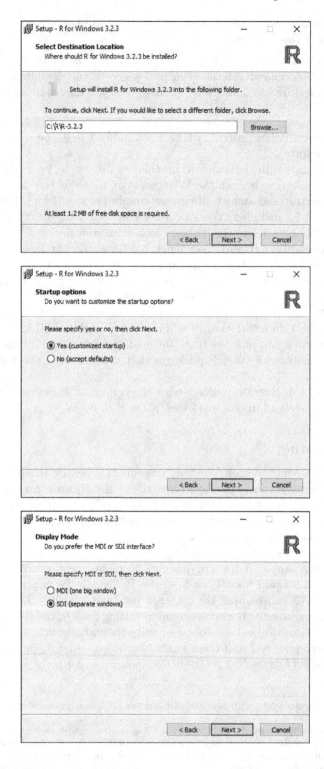

FIGURE 2.2: Key steps in installing R for Windows (illustrated with the R for Windows version 3.2.3 installer): installing R into the directory C:\R\R-3.2.3 (top); selecting customized startup options (middle); selecting the SDI (bottom).

If, however, you inadvertently saved the workspace in a previous session, it will reside in a file named `.RData`. To discover where this file is located, enter the command `getwd()` ("get working directory") at the R > command prompt as soon as R starts up.

If Windows is configured to suppress known file types (also called file extensions), as it is by default, then you will not see the name of this file in the Windows File Explorer, because the file name begins with a period (`.`) and thus, as far as Windows is concerned, consists *only* of a file type. You can, however, still delete the file by right-clicking the R icon with no file name that appears at the top of the alphabetized *Name* column in the File Explorer window, selecting *Delete* from the resulting context menu. Be careful not to delete any other (named) files associated with the R icon.

2.3 Installing **R** and the **R Commander** under **Mac OS X**

Before installing R, make sure that your Mac OS X system is up-to-date by running *Software Update* from the "Apple" menu at the top-left of the screen. This is important, because R assumes that your system is up-to-date and may not function properly if it is not.

From the home page of the CRAN mirror you selected, click on the link *Download R for (Mac) OS X*, which appears near the top of the page; then click on *R-x.y.z.pkg*.[4] Once it is downloaded, double-click on the R installer. You may take all of the defaults.

The initial screen of the R for Mac OS X installer is shown at the top of Figure 2.3.

2.3.1 Installing **X11** (**XQuartz**) for **Mac OS X**

Background: The R Commander uses the **tcltk** package, which is a standard component of the R distribution. On Mac OS X systems, R also installs a version of the Tcl/Tk GUI builder, used by the **tcltk** package. This version of Tcl/Tk in turn uses the Unix X11 windowing system (also called X Windows) instead of the standard Mac Quartz windowing system.

Some older versions of Mac OS X had X11 pre-installed, while other older versions came with X11 on the operating system installation disks. X11 is absent from newer versions of Mac OS X, but is readily available on the Internet at the XQuartz web site, http://xquartz.macosforge.org.[5]

I suggest that you simply install the current version of XQuartz, whether or not an older version of X11 is installed on your computer:

- Download the disk-image file (`XQuartz-x.y.z.dmg`) for the current version *x.y.z* of XQuartz.

- When you open this file by double-clicking on it, you'll find `XQuartz.pkg`; double-click on `XQuartz.pkg` to run the XQuartz installer, clicking through all the defaults. The initial screen of the XQuartz installer is shown at the bottom of Figure 2.3.

- After the installer runs, you'll have to log out of and back into your Mac OS X account—or just reboot your computer. You can remove the XQuartz disk image from your desktop by dragging it and dropping it on the Mac OS X trash can.

---

[4]If you have an older version of Mac OS X, you may not be able to use the current version of R, but an older version of R compatible with your operating system may be provided: Read the information on the *R for Mac OS X* web page before downloading the R installer.

[5]Although it is potentially confusingly named, XQuartz is an implementation of the X11 windowing system for Mac OS X.

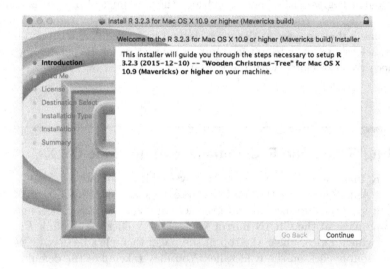

FIGURE 2.3: The initial screens of the R for Mac OS X installer (version 3.2.3, top) and the XQuartz installer (version 2.7.8, bottom).

If you subsequently upgrade Mac OS X (e.g., from version 10.10 to version 10.11), you will have to reinstall XQuartz (and possibly R itself), even if you installed it previously.

2.3.2 Installing the Rcmdr Package and Its Dependencies

Once R and X11 are installed, start R in the standard manner—for example, by double-clicking on the R icon in the `Applications` folder.

The easiest way to install the **Rcmdr** package, if you have an active Internet connection, is via the *Packages & Data* menu in the *R Console*: Click on the *Packages & Data* menu and select the *Package Installer* menu item.

- Type Rcmdr in the *Package Search* box, and click the *Get List* button.

- R will ask you to select a CRAN mirror; as before, I suggest that you pick the first "0-Cloud" mirror, and that, when asked, you opt to set the selected mirror as the default.

- Click on the **Rcmdr** package in the resulting packages list; check the *Install Dependencies* box, and click on the *Install Selected* button.

- Once the **Rcmdr** package and its dependencies are installed, which may take a bit of time, you can close the *R Package Installer* window.

An alternative to using the R menus is to type `install.packages("Rcmdr")` at the > command prompt in the *R Console* (followed by pressing the *return* or *enter* key). R will install the **Rcmdr** package, along with a number of other R packages that the R Commander requires to get started.

When you first load the **Rcmdr** package with the command `library(Rcmdr)`, it may offer to download and install additional packages; if so, allow it to do so.

On some versions of Mac OS X, you may see an additional message from R the first time that you load the **Rcmdr** package: "The 'otool' command requires the command line developer tools. Would you like to install the tools now?" If you see this message, click the *Install* button in the message dialog box.

2.3.3 Preventing R from Napping

Under Mac OS X 10.9 ("Mavericks") or later, the R Commander may slow down or occasionally hesitate to display a menu as your session progresses. This behavior is due to Mac OS X saving power by going into "nap" mode (called app nap) when the *R.app* window is not visible.

I am aware of several solutions (beyond inconveniently insuring that the top of the *R.app* window is always visible). The simplest solution is to suppress app nap via the R Commander menus: *Tools > Manage Mac OS X app nap for R.app*. That is, choose the menu item *Manage Mac OS X app nap for R.app* from the R Commander *Tools* menu. In the resulting dialog, click the radio button to set app nap off. This setting is permanent across R.app sessions until you change it.

For alternative solutions, see the web site for the book.

2.3.4 Troubleshooting

Occasionally, the **Rcmdr** package will fail to load properly in Mac OS X. When this problem occurs, the cause is almost always the failure of the **tcltk** package to load. The problem is usually clearly stated in an error message printed in the R console. You can confirm the diagnosis by trying to load the **tcltk** package directly in a fresh R session, issuing the command `library(tcltk)` at the R command prompt.

The solution is almost always to install, or reinstall, XQuartz (and possibly R), as described above, remembering to log out of and back into your account before trying to run R and the R Commander again. If this solution fails, then you can consult the more detailed troubleshooting information in the Mac OS X installation notes on the web site for the book.

Beyond the failure of the **tcltk** package to load, occasionally, and unpredictably, an R package required by the **Rcmdr** package fails to be installed—possibly because the package is missing from the CRAN mirror that you used—and the R Commander can't start. Under these circumstances, there is typically an informative error message about the missing package.

The simple solution is to install the missing package (or packages, if there are more than one) directly. For example, if the **car** package is missing, you can install it via the R command `install.packages("car")`, or from the *R.app Packages* menu, possibly selecting a different CRAN mirror from the one that you used initially.

Sometimes when users save the R workspace upon exiting from R, the R Commander will fail to work properly in a subsequent session. As I will explain in Section 3.8, I recommend *never* saving the R workspace to avoid these kinds of problems, and if you exit from R via the R Commander menus (*File > Exit > From Commander and R*), the R workspace will not be saved.

If, however, you inadvertently saved the workspace in a previous session, it will reside in a file named `.RData`. To discover where this file is located, enter the command `getwd()` ("get working directory") at the R > command prompt as soon as R.app starts up; this will typically be your home directory.

Unfortunately, newer versions of Mac OS X don't make it easy for you to view the contents of your home directory in the Finder. Instead, run the Mac OS X Terminal program; you'll find Terminal in the Mac OS X Utilities subfolder within the Applications folder. Type the command `ls -a` at the Terminal $ command prompt (followed by pressing the *enter* or *return* key) to list all files in your home directory. Among these files, you should see `.RData`. Then type the command `rm .Rdata` to remove the offending file.

2.4   Installing **R** and the **R Commander** on **Linux** and **Unix** Computers

R is available from CRAN for several Linux distributions (Debian, RedHat, SUSE, and Ubuntu); select your distribution, and proceed as directed.

If you have a Linux or Unix system that's not compatible with one of these distributions, then you will have to compile R from source code; the procedure for doing so is described in the R FAQ ("frequently asked questions") list at https://cran.r-project.org/doc/FAQ/R-FAQ.html (Question 2.5.1, at the time of writing).

Once R is installed, you will have to install the **Rcmdr** package and its dependencies. Start R and type the command `install.packages("Rcmdr")` at the > command prompt (followed by pressing the *Return* or *Enter* key). You may be asked to select a CRAN mirror site; as before, I suggest that you pick the first "0-Cloud" mirror. After the **Rcmdr** and its direct package dependencies are installed, start the R Commander via the command `library(Rcmdr)`. The R Commander will ask to install some additional packages; let it do that.

2.4.1 Troubleshooting

Occasionally and unpredictably, an R package required by the **Rcmdr** package fails to be installed—possibly because the package is missing from the CRAN mirror that you used—and the R Commander can't start. Under these circumstances, there is typically an informative error message about the missing package.

The simple solution is to install the missing package (or packages, if there are more than one) directly. For example, if the **car** package is missing, you can install it via the R command `install.packages("car")`, possibly selecting a different CRAN mirror from the one that you used initially.

Sometimes when users save the R workspace upon exiting from R, the R Commander will fail to work properly in a subsequent session. As I will explain in Section 3.8, I recommend *never* saving the R workspace to avoid these kinds of problems, and if you exit from R via the R Commander *File > Exit > From Commander and R* menu, the R workspace will not be saved.

If, however, you inadvertently saved the workspace in a previous session, it will reside in a file named `.RData`. To discover where this file is located, enter the command `getwd()` ("get working directory") at the R > command prompt as soon as R starts up; this will typically be your home directory.

Type the command `ls -a` at the command prompt in a fresh Linux terminal to list all files in your home directory. Among these files, you should see `.RData`. Then type the command `rm .Rdata` to remove the offending file.

You may also find that you are missing a C or Fortran compiler, required by R to build packages, or an installation of Tcl/Tk, required by the **tcltk** package, which is in turn used by the R Commander. If you experience these or other difficulties, consult the R FAQ ("frequently asked questions") at https://cran.r-project.org/doc/FAQ/R-FAQ.html, and the *R Installation and Administration* manual at https://cran.r-project.org/doc/manuals/r-release/R-admin.html (particularly Appendix A on essential and useful programs).

2.5 Installing Optional Auxiliary Software: **Pandoc** and LᴬTᴇX

Installing R and the **Rcmdr** package is sufficient for creating HTML (web page) reports (see Section 3.6), but if you prefer to create editable Word documents or PDF files for reports, you must additionally install Pandoc and LᴬTᴇX (the latter, in conjunction with Pandoc, is needed only for PDF reports).[6] The most convenient way to do this is via the R Commander menus: *Tools > Install auxiliary software.*[7]

---

[6]Pandoc is a flexible program for converting documents from one format to another, while LᴬTᴇX is technical typesetting software—this book, for example, is typeset with LᴬTᴇX. Both Pandoc and LᴬTᴇX are open-source software.

[7]The *Install auxiliary software* menu item appears in the *Tools* menu only if Pandoc or LᴬTᴇX is missing.

3

A Quick Tour of the R Commander

This chapter introduces the R Commander graphical user interface (GUI) by demonstrating its use for a simple problem: constructing a contingency table to examine the relationship between two categorical variables. In developing the example, I explain how to start the R Commander, describe the structure of the R Commander interface, show how to read data into the R Commander, how to modify data to prepare them for analysis, how to draw a graph, how to compute numerical summaries of data, how to create a printed report of your work, how to edit and re-execute commands generated by the R Commander, and how to terminate your R and R Commander session—in short, the typical work flow of data analysis using the R Commander. I also explain how to customize the R Commander interface.

In the course of this chapter, you'll get an overview of the operation of the R Commander. Later in the book, I'll return in more detail to many of the topics addressed in the chapter.

3.1   Starting the **R Commander**

I assume that you have installed R and the **Rcmdr** package, as described in the preceding chapter. As well, if you haven't read Chapter 1, now is a good time to do so—Chapter 1 explains some typographical conventions used in this book, discusses the general characteristics and origin of R and the R Commander, and introduces the web site for the book.

Start R in the normal manner for your computer, for example, by double-clicking on the R desktop icon in Windows, by double-clicking on R.app in the Mac OS X *Applications* folder, or by clicking on the R icon in the Mac OS X *Launchpad.*[1] On a Linux or Unix machine, you'd normally start R by typing R at the command prompt in a terminal window.

Once R starts up, type the command library(Rcmdr) at the R > command prompt, and then press the *Enter* or *Return* key. This command should load the **Rcmdr** package and—after a brief delay—start the R Commander GUI, as shown in Figure 3.1 for Windows[2] or Figure 3.2 for Mac OS X. If you encounter a problem in starting R or the R Commander, see the sections on troubleshooting in Chapter 2 (Section 2.2.1 for Windows, 2.3.4 for Mac OS X, or 2.4.1 for Linux/Unix).

---

[1]If you plan to use R and the R Commander frequently under Mac OS X, it is convenient to add the R icon to the dock.

[2]This is how the *R Console* appears in the R for Windows single-document interface (SDI) that I recommended in the installation instructions in Chapter 2. If you instead installed R with the default multiple-document interface (MDI), then the *R Console* appears inside a larger *RGui* window—not an ideal arrangement for the R Commander.

3.2   The **R Commander** Interface

Under Windows, the R Commander (Figure 3.1) looks like a standard program. In contrast, under Mac OS X (Figure 3.2), the R Commander has its own main *menu bar*, unlike a standard application, which would use the menu bar at the top of the Mac OS X desktop.[3]

As you can see, the main *R Commander* window looks very similar under Windows and Mac OS X. After this introductory chapter, I will show R Commander dialog boxes as they appear under Windows 10. As well, all dialogs and graphs in the text are rendered in monochrome (gray-scale) rather than in color.[4]

At the top of the *R Commander* window there is a menu bar with the following top-level menus:

File contains menu items for opening and saving various kinds of files, and for changing the R *working directory*—the folder or directory in your file system where R will look for and write files by default.

Edit contains common menu items for editing text, such as *Copy* and *Paste*, along with specialized items for R Markdown documents (discussed in Section 3.6.2).

Data contains menu items and submenus for importing, exporting, and manipulating data (see in particular Sections 3.3 and 3.4, and Chapter 4).

Statistics contains submenus for various kinds of statistical data analysis (discussed in several subsequent chapters), including fitting statistical models to data (Chapter 7).

Graphs contains menu items and submenus for creating common statistical graphs (see in particular Chapter 5).

Models contains menu items and submenus for performing various operations on statistical models that have been fit to data (see Chapter 7).

Distributions contains a menu item for setting the R random-number-generator seed for simulations, and submenus for computing, graphing, and sampling from a variety of common (and not so common) statistical distributions (see Chapter 8).

Tools contains menu items for loading R packages and R Commander plug-in packages (see Chapter 9), for setting and saving R Commander options (see Section 3.9), for installing optional auxiliary software (see Section 2.5), and, under Mac OS X, for managing app nap for R.app (see Section 2.3.3).

Help contains menu items for obtaining information about the R Commander and R, including links to a brief introductory manual and to the R Commander and R web sites; information about the active data set; and a link to a web site with detailed instructions for using R Markdown to create reports (see Section 3.6).

The complete R Commander menu tree is shown in the appendix to this book (starting on page 199).

---

[3]As explained in Chapter 2, the Tcl/Tk GUI builder installed with R and used by the R Commander employs the X11 windowing system rather than the native Mac Quartz windowing system. This is why the R Commander can't use the standard Mac OS X top-level menu bar.

[4]In the few instances in which color is important to the interpretation of a figure, the figure is repeated in its original color version in the center insert to the book; these instances are noted in the figure captions.

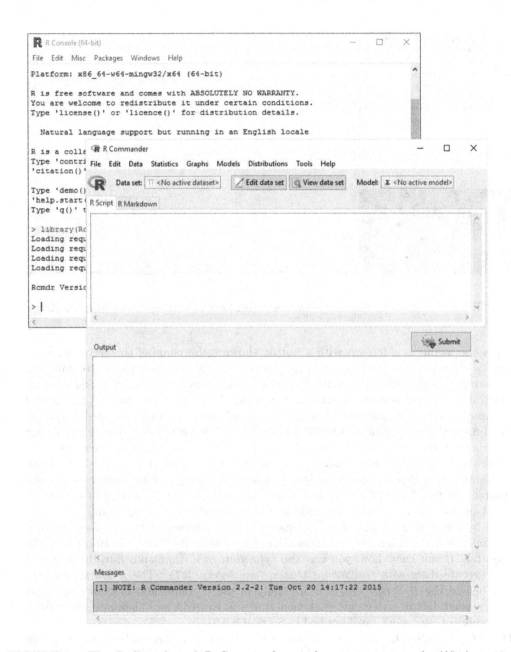

FIGURE 3.1: The *R Console* and *R Commander* windows at startup under Windows 10.

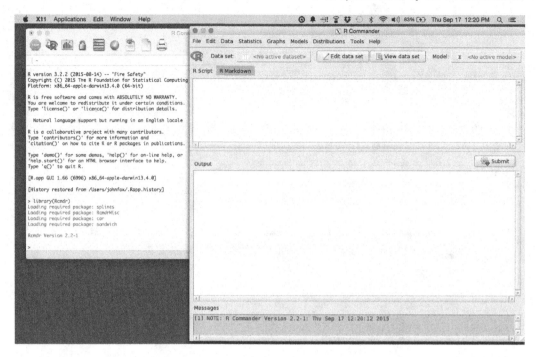

FIGURE 3.2: The *R.app* and *R Commander* windows at startup under Mac OS X.

Below the menus is a *toolbar*, with a button showing the name of the active data set (displaying <*No active dataset*> at startup), buttons to edit and view the active data set, and a button showing the active statistical model (displaying <*No active model*> before a statistical model has been fit to data in the active data set). The *Data set* and *Model* buttons may also be used to choose from among multiple data sets and associated statistical models if more than one data set or model resides in the R *workspace*—the region of your computer's main memory where R stores data sets, statistical models, and other objects.

Below the toolbar there is a window pane with two tabs, labelled respectively *R Script* and *R Markdown*, that collect the R commands generated during your R Commander session. The contents of the *R Script* and *R Markdown* tabs can be edited, saved, and reused (as described in Section 3.6), and commands in the *R Script* tab can be modified and re-executed by selecting a command or commands with the mouse (left-click and drag the mouse cursor over the command or commands) and pressing the *Submit* button below the *R Script* tab. If you know how, you can also type your own commands into the *R Script* tab and execute them with the *Submit* button (see Section 3.7).[5] The *R Markdown* tab, initially behind the *R Script* tab, also accumulates the R commands that are generated during a session, but in a dynamic document that you can edit and elaborate to create a printed report of your work (as described in Section 3.6.2).

The R Commander *Output* pane appears next: The *Output* pane collects R commands generated by the R Commander along with associated printed output. The text in the *Output* pane is also editable, and it can be copied and pasted into other programs (as described in Section 3.6.1).

---

[5]If a command is self-contained on a single line, then it can be executed by pressing *Submit* when the cursor is anywhere in the line; if a command extends over several lines, however, then all lines must be selected and submitted simultaneously for execution.

Finally, at the bottom of the R Commander window, the *Messages* pane records messages generated by R and the R Commander—numbered and color-coded *notes* (dark blue), *warnings* (green), and *error messages* (red). For example, the startup note indicates the R Commander version, along with the date and time at the start of the session.

Once you have started the R Commander GUI, you can safely minimize the *R Console* window—this window occasionally reports messages, such as when the R Commander causes other R packages to be loaded, but these messages are incidental to the use of the R Commander and can almost always be safely ignored.[6]

3.3 Reading Data into the R Commander

Statistical data analysis in the R Commander is based on an *active data set* in the form of an R *data frame*. A data frame is a *rectangular data set* in which the rows (running horizontally) represent *cases* (often individuals) and the columns (running vertically) represent *variables* descriptive of those cases. Columns in data frames can contain various forms of data—*numeric variables*, *character-string variables* (with values such as "Yes", "No", or "Maybe"), *logical variables* (with values TRUE or FALSE), and *factors*, which are the standard representation of categorical data in R. Typically, data frames used in the R Commander consist of numeric variables and factors, and character and logical variables, if present, are treated as factors.

R and the R Commander permit you to have as many data frames in your workspace as will fit,[7] but only one is active at any given time. You can read data into data frames from several sources using the R Commander menus:[8] See the *Data > Import data* submenu, and the *Data > Data in packages > Read data set from an attached package* menu item and associated dialog. If more than one data frame resides in your workspace, you can choose among them by pressing the *Data set* button in the toolbar or via the menus: *Data > Active data set > Select active data set*.

One convenient source of data is a *plain-text* (*"ASCII"*) file with one line per case, variable names in the first line, and values in each line separated by a simple delimiter such as spaces or a comma. An example of a plain-text data file with *comma-separated values*, GSS.csv, is shown in Figure 3.3.[9]

The data in the file GSS.csv are drawn from the U.S. General Social Survey (GSS), and were collected between 1972 and 2012. The GSS is a periodic cross-sectional sample survey of the U.S. population conducted by the National Opinion Research Center at the University of Chicago. Many of the questions in the GSS are repeated in each survey, while other questions are repeated at intervals. To compile the GSS data set, I selected instances of the GSS that asked the question, "There's been a lot of discussion about the way morals and attitudes about sex are changing in this country. If a man and a woman have sex relations before marriage, do you think it is always wrong, almost always wrong, wrong only sometimes, or not wrong at all?" I also included information about the year of the

---

[6]Before minimizing the Mac OS X *R Console*, however, make sure that app nap is turned off, or—as explained in Section 2.3.3—the R Commander may become unresponsive!

[7]Unless you're working with massive data sets, in which case the R Commander is probably not a good choice of interface to R, fitting data into the R workspace will not be an issue.

[8]Data input from various sources is discussed in more detail in Chapter 4.

[9]GSS.csv and other data files employed in this book are available for download on the web site for the book: See Section 1.5.

```
year,gender,premarital.sex,education,religion
1972,female,not wrong at all,post-secondary,Jewish
1972,male,always wrong,less than high school,Catholic
1972,female,always wrong,high school,Protestant
1972,female,always wrong,post-secondary,other
1972,female,sometimes wrong,high school,Protestant
 . . .
2012,female,not wrong at all,post-secondary,none
2012,male,not wrong at all,high school,Catholic
2012,female,sometimes wrong,high school,Catholic
```

FIGURE 3.3: The GSS.csv file, with comma-delimited data from the U.S. General Social Survey from 1972 to 2012. Only a few of the 33,355 lines in the file are shown; the widely spaced ellipses (. . .) represent elided lines. The first line in the file contains variable names.

TABLE 3.1: Variables in the GSS data set.

Variable	Values
year	numeric, year of survey, between 1972 and 2012
gender	character, female or male
premarital.sex	character, always wrong, almost always wrong, sometimes wrong, or not wrong at all
education	character, less than high school, high school, or post-secondary
religion	character, Protestant, Catholic, Jewish, other, or none

survey, and the respondents' gender, education, and religion. Table 3.1 shows the definition of the variables in the GSS data set.

This is a natural point at which to explain how objects, including data sets and variables, are named in R: Standard R names are composed of lower- and upper-case letters (a–z, A–Z), numerals (0–9), periods (.), and underscores (_), and must begin with a letter or a period. As well, R is case sensitive; so, for example, the names education, Education, and EDUCATION are all distinct.

In order to keep this introductory example as simple as possible, when I compiled the GSS data set from the original source, I eliminated cases with missing values for any of the four substantive variables (of course, there were no missing values for the year of the survey). In R, missing values are represented by NA ("not available"), and in the R Commander, NA is the default missing-data code for text-data input, although another missing-data code (such as ?, ., or 99) can be specified. This and some other complications and variations are discussed in Chapter 4 on reading and manipulating data in the R Commander.

To read simply formatted data in plain-text files into the R Commander, you can use *Data > Import data > from text file, clipboard, or URL*. As the name of this menu item implies, the data can be copied to the clipboard (e.g., from a suitably formatted spreadsheet) or read from a file on the Internet, but most often the data will reside in a file on your computer.

The resulting dialog box is shown in Figure 3.4. This is a comparatively simple R Commander dialog box—for example, it doesn't have multiple tabs—but it nevertheless illustrates several common elements of R Commander dialogs:

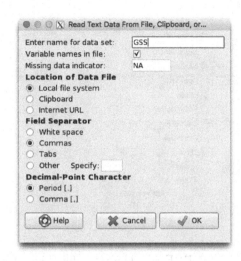

FIGURE 3.4: The *Read Text Data* dialog as it appears on a Windows computer (left) and under Mac OS X (right).

- There is a *check box* to indicate whether variable names are included with the data, as they are in the GSS.csv data file.

- There are *radio buttons* for selecting one of several choices—here, where the data are located, how data values are separated, and what character is used for decimal points (e.g., commas are used in France and the Canadian province of Québec).

- There are *text fields* into which the user can type information—here, the name of the data set, the missing-data indicator, and possibly the data-field separator.

I've taken all of the defaults in this dialog box, with the following two exceptions: I changed the default data set name, which is Dataset, to the more descriptive GSS. Recall the rules, explained above, for naming R objects. For example, GSS data, with an embedded blank, would not be a legal data set name. I also changed the default field separator from *White space* (one or more spaces or a tab) to *Commas*, as is appropriate for the comma-separated-values file GSS.csv.

The *Read Text Data* dialog also has buttons at the bottom that are standard in R Commander dialogs:

- The *Help* button opens an R *help page* in a web browser, documenting either the use of the dialog or the use of an R command that the dialog invokes. In this case, pressing the *Help* button opens up the help page for the R read.table function, which is used to input simple plain-text data. R help pages are *hyper-linked*, so clicking on a link will open another, related help page in your browser. (Try it!)

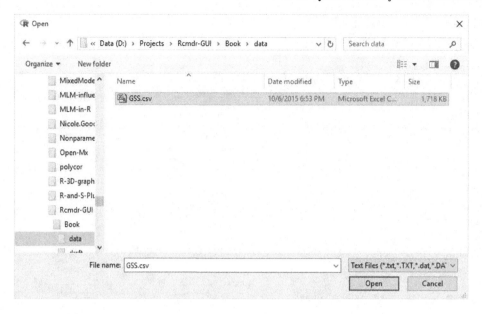

FIGURE 3.5: The *Open* file dialog with the data file `GSS.csv` selected.

- Pressing the *OK* button generates and executes an R command (or, in the case of some dialogs, a sequence of R commands).[10] These commands are usually entered into the *R Script* and *R Markdown* tabs, and the commands and associated printed output appear in the *Output* pane. If graphical output is produced, it appears in a separate R *graphics-device* window.

  Clicking *OK* in the *Read Text Data* dialog brings up a standard *Open* file dialog box, as shown in Figure 3.5. I navigated to the location of the data file on my computer and selected the `GSS.csv` file. Notice that files of type `.csv`, `.txt`, and `.dat` (and their upper-case analogs) are listed by default—these are common file types associated with plain-text data files.

  Clicking *OK* causes the data to be read from `GSS.csv`, creating the data frame `GSS`, and making it the active data set in the R Commander. The `read.table` command invoked by the dialog converts character data in the input file to R factors (here, the variables `gender`, `premarital.sex`, `education`, and `religion`).

- Clicking the *Cancel* button simply dismisses the *Read Text Data* dialog.

As is apparent, the order of the buttons at the bottom of the dialog box is different in Windows and in Mac OS X, reflecting differing GUI conventions on these two computing platforms.

---

[10]*OK* is also the *default button* in the dialog (under Windows, the button is outlined in blue), and so pressing the *Return* or *Enter* key is equivalent to left-clicking on the button.

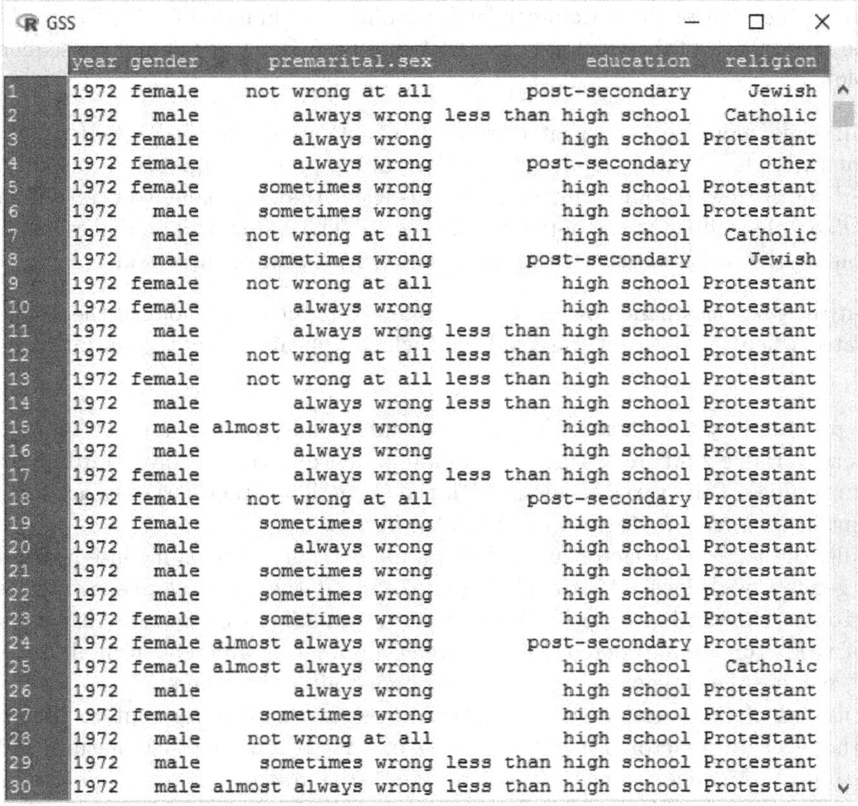

	year	gender	premarital.sex	education	religion
1	1972	female	not wrong at all	post-secondary	Jewish
2	1972	male	always wrong	less than high school	Catholic
3	1972	female	always wrong	high school	Protestant
4	1972	female	always wrong	post-secondary	other
5	1972	female	sometimes wrong	high school	Protestant
6	1972	male	sometimes wrong	high school	Protestant
7	1972	male	not wrong at all	high school	Catholic
8	1972	male	sometimes wrong	post-secondary	Jewish
9	1972	female	not wrong at all	high school	Protestant
10	1972	female	always wrong	high school	Protestant
11	1972	male	always wrong	less than high school	Protestant
12	1972	male	not wrong at all	less than high school	Protestant
13	1972	female	not wrong at all	less than high school	Protestant
14	1972	male	always wrong	less than high school	Protestant
15	1972	male	almost always wrong	high school	Protestant
16	1972	male	always wrong	high school	Protestant
17	1972	female	always wrong	less than high school	Protestant
18	1972	female	not wrong at all	post-secondary	Protestant
19	1972	female	sometimes wrong	high school	Protestant
20	1972	male	always wrong	high school	Protestant
21	1972	male	sometimes wrong	high school	Protestant
22	1972	male	sometimes wrong	high school	Protestant
23	1972	female	sometimes wrong	high school	Protestant
24	1972	female	almost always wrong	post-secondary	Protestant
25	1972	female	almost always wrong	high school	Catholic
26	1972	male	always wrong	high school	Protestant
27	1972	female	sometimes wrong	high school	Protestant
28	1972	male	not wrong at all	high school	Protestant
29	1972	male	sometimes wrong	less than high school	Protestant
30	1972	male	almost always wrong	less than high school	Protestant

FIGURE 3.6: The R Commander data-set viewer displaying the `GSS` data set.

3.4 Examining and Recoding Variables

Having read data into the R Commander from an external source, it's generally a good idea to take a quick look at the data, if only to confirm that they've been read properly. Clicking the *View data set* button in the R Commander toolbar brings up the data-viewer window shown in Figure 3.6. Variable names remain at the top of the display as the rows are scrolled using the scrollbar at the right of the data-viewer window. Row numbers appear to the left of the data; if the rows of the data set were named, the row names would appear here (and row numbers or names remain at the left if it's necessary to scroll the data viewer horizontally). You may leave the data-viewer window open on your desktop as you continue to work in the R Commander, or you may close the data viewer. If you leave it open, the data viewer will be automatically updated if you make subsequent changes to the active data set.

Although the `GSS` data set contains a moderately large number of cases (with $n = 33,354$ rows), there are only five variables, and so I request a summary of all the variables in the data set, invoked by *Statistics* > *Summaries* > *Active data set*. The result is shown in Figure 3.7:

- R commands generated in the R Commander session are accumulated in the *R Script* tab (and in the *R Markdown* tab, which is currently behind the *R Script* tab and consequently isn't visible).

- These commands, along with associated printed output, appear in the *Output* pane; the scrollbar at the right of the pane allows you to examine previous input and output that has scrolled out of view. If some printed material is wider than the pane, you can similarly use the horizontal scrollbar at the bottom to inspect it. The R Commander makes an effort to fit output to the width of the *Output* pane, but it isn't always successful.

- Notice that the *Messages* pane now includes a note about the dimensions of the GSS data set, generated when the data set was read, and which appears below the initial start-up message.

The output produced by the summary(GSS) command includes a "five-number summary" for the numeric variable year, reporting the minimum, first quartile, median, third quartile, and maximum values of the variable, along with the mean. The other variables are factors, and the count in each *level* (category) of the factor is shown.

By default, the levels of a factor are ordered alphabetically. This is inconsequential in the case of gender, with levels "female" and "male", but the levels of premarital.sex and education have natural orderings different from the alphabetic orderings. Although the categories of religion are unordered, I'd still prefer an ordering different from alphabetic, for example, putting the categories "other" and "none" after the others.

I won't use all of the variables in the GSS data set in this chapter, but to illustrate reordering the levels of a factor, I'll put the levels of education into their natural order. Clicking on *Data > Manage variables in active data set > Reorder factor levels* produces the dialog box at the left of Figure 3.8. I select education in the variable list box in the dialog, leave the name for the factor at its default value <same as original>, and keep the *Make ordered factor* box unchecked.[11] Because the variable name is unchanged, the new education variable will *replace* the original variable in the GSS data frame, and so the R Commander will ask for confirmation when I click the *OK* button.

Variable list boxes are a common feature of R Commander dialogs:

- In general, left-clicking on a variable in an R Commander variable list selects the variable.

- If more than one variable is to be selected—which is *not* the case in the *Reorder Factor Levels* dialog—you can *Ctrl*-left-click to choose additional variables—that is, simultaneously hold down the *Ctrl* (*Control*) key on your keyboard and click the left mouse button.

- *Ctrl*-clicking "toggles" a selection, so if a variable is *already* selected, *Ctrl*-clicking on its name will de-select it.

- The *Ctrl* key is used in the same way on Macs and on PCs, although on a Mac keyboard, the key is named *control*. You *cannot* use the Mac *command* key here instead of *control*.

- Similarly, *Shift*-clicking may be used to select a contiguous range of variables in a list: Click on a variable at one end of the desired range and then *Shift*-click on the variable at the other end.

- Finally, you can use the scrollbar in a variable list if the list is too long to show all of the variables simultaneously, and pressing a letter key scrolls to the first variable whose name

---

[11]An *ordered factor* in R is a factor whose levels are recognized to have an intrinsic ordering. I *could* use an ordered factor here, but there is no real advantage to doing so, and I will not employ ordered factors in this book.

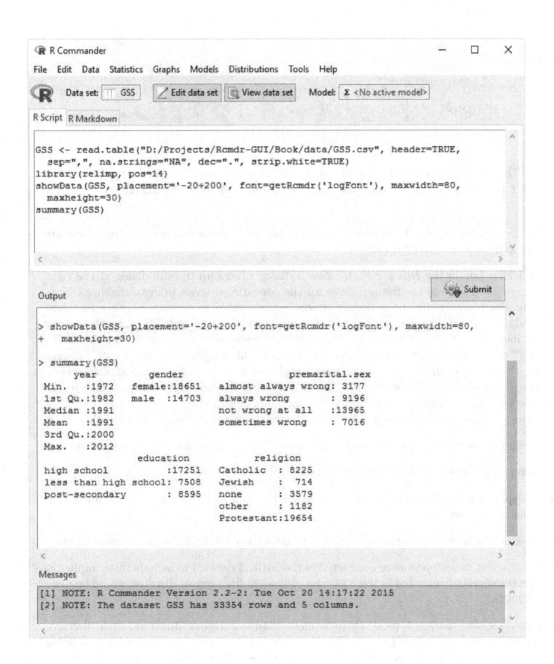

FIGURE 3.7: The *R Commander* window after summarizing the active data set.

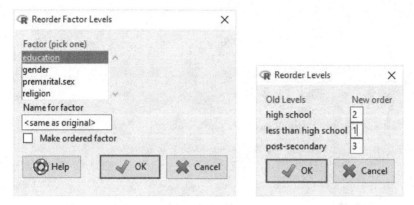

FIGURE 3.8: The *Reorder Factor Levels* dialog with `education` selected (left), and the *Reorder Levels* sub-dialog showing the reordered levels of `education` (right).

begins with that letter. It's unnecessary to scroll the variable list here because there are only four factors in the data set.

Clicking *OK* in the *Reorder Factor Levels* dialog brings up the sub-dialog at the right of Figure 3.8. I type in the natural order for the educational levels prior to clicking *OK*.

My aim is eventually to construct a contingency table to explore whether and how attitude towards premarital sex has changed over time. To this end, I'll group the survey years into decades. As well, because relatively few respondents answered "almost always wrong" to the premarital-sex question, I'll combine this response category with "always wrong." Both operations can be performed with the *Recode* dialog, invoked via *Data > Manage variables in active data set > Recode variables*. The resulting dialog, completed to recode `year` into `decade`, is displayed in Figure 3.9.

The following syntax is employed in the *Enter recode directives* box in the dialog:

- A colon (`:`) is used to specify a range of values of the original numeric variable `year`.

- Factor levels (such as `"1970s"`) are enclosed in double quotes.

- The special values `lo` and `hi` can be used for the minimum and maximum values of a numeric variable.

- An equals sign (`=`) associates each set of old values with a level of the factor to be created.

- Because just two surveys were conducted in the 2010s, I decided to include these implicitly with the surveys conducted in the 2000s; an equivalent final recode directive would be `else = "2000s"`.

- When, as here, and is typical, there is more than one recode directive, each directive appears on a separate line; press the *Enter* or *return* key on your keyboard when you finish typing each recode directive to move to the next line before typing the next directive.

- Press the *Help* button (and read Section 4.4.1) to see more generally how recode directives are specified.

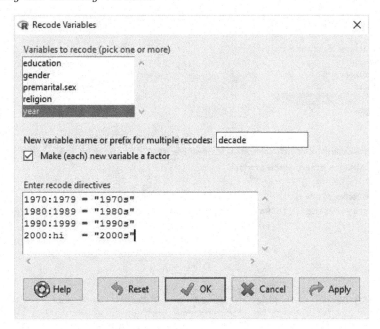

FIGURE 3.9: The *Recode Variables* dialog, recoding `year` into `decade`.

In this example, I select the variable `year` from the variable list. I replace the default variable name (which is `variable`) with `decade`. I also leave the box checked to make `decade` a factor.[12]

In addition to the now-familiar *Help*, *OK*, and *Cancel* buttons, there are also *Apply* and *Reset* buttons in the *Recode Variables* dialog:

- Clicking the *Apply* button is like clicking *OK*, except that, after generating and executing a command or set of commands, the dialog reopens in its previous state.

- As a general matter, R Commander dialogs "remember" their state from one invocation to the next if that's sensible—for example, if the active data set hasn't changed. Pressing the *Reset* button in a dialog restores the dialog to its pristine state.

After clicking the *Apply* button, the *Recode Variables* dialog reappears in its previous state. Because `premarital.sex` is to be recoded entirely differently from `year`, I then press the *Reset* button, and specify the desired recode, shown in Figure 3.10. I select `premarital.sex` as the variable to be recoded, and enter `premarital` as the name of the new factor to be created; I could have used the same name as the original variable, in which case the R Commander would have asked for confirmation. Because I intend to leave the levels `sometimes wrong` and `not wrong at all` alone, I don't have to recode them.

The single recode directive in the dialog changes `"almost always wrong"` and `"always wrong"` to `"wrong"`, with the values of the original factor `premarital.sex` on the left of `=` separated by a comma. This recode directive is too long to appear in its entirety in the dialog box, but the scroll bar at the bottom of the *Enter recode directives* text box allows you to see it; the whole directive is `"almost always wrong", "always wrong" = "wrong"`. Because

---

[12]More than one variable can be selected if the same recode directives are to be applied to each. If several variables are to be recoded, the name supplied is treated as a *prefix* for the new variable names, which are formed by pasting the prefix onto the names of the recoded variables.

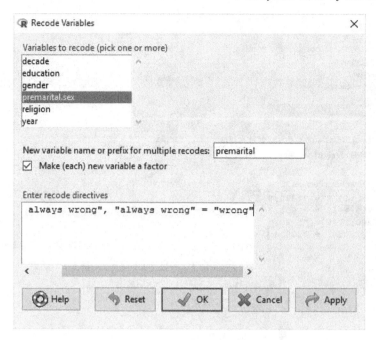

FIGURE 3.10: The *Recode Variables* dialog, recoding `premarital.sex` into `premarital`. Because of its length, the recode directive `"almost always wrong", "always wrong" = "wrong"` isn't entirely visible and the scrollbar below the *Enter recode directives* box is activated.

the levels of the new factor `premarital` are already in their natural order (alphabetically, `not wrong at all`, `sometimes wrong`, `wrong`), I don't have to reorder them subsequently.

An advantage of a graphical user interface like the R Commander is that choices usually are made by pointing and clicking, minimizing the necessity to type and thus tending to reduce errors. The R Commander doesn't *entirely* eliminate typing, however: You must be careful when typing recode directives, and more generally in the R Commander when you type text into a dialog box. If, for example, you type an existing factor level incorrectly in a recode directive, the directive will have no effect. You have to include spaces, and other punctuation such as commas, if these appear in level names. Also remember that R names are case sensitive.

Notice that I've used the *Recode* dialog for two distinct purposes:

1. I created a factor (`decade`) from a numeric variable (`year`) by dissecting the range of the numeric variable into class intervals, often called *bins*. Binning is useful because it allows me to make a contingency table (in Section 3.5), relating attitude towards premarital sex to the date of the survey; there are too many distinct values of `year` to treat them as separate categories in the contingency table, and in the extreme case of a truly continuous numeric variable, all of the values of the variable may be unique.

2. I combined some categories of a factor (`premarital.sex`) to create a new factor (`premartial`). Combining factor categories was useful because one of the levels of `premarital.sex`, `"almost always wrong"`, was chosen by relatively few respondents.

FIGURE 3.11: The *Bar Graph* dialog, showing the *Data* and *Options* tabs. The *x-axis label*, `Attitude Towards Premarital Sex`, is too long to be visible in its entirety, so the scrollbar below the label is activated.

I could examine the distribution of the recoded variable by selecting *Statistics > Summaries > Frequency distributions*, picking `premarital` in the resulting dialog, but— primarily to illustrate drawing a graph—I instead construct a bar graph of the distribution: Selecting *Graphs > Bar graph* produces the dialog box in Figure 3.11. The same dialog in Mac OS X is shown in Figure 3.12.

As is common in R Commander dialogs, there are two tabs in the *Bar Graph* dialog box, in this case named *Data* and *Options*. I select `premarital` from the variable list in the *Data* tab. Were I to click the *Plot by groups* button, a sub-dialog box would open, permitting me to select a grouping factor, with one bar graph constructed for each group. I type `Attitude towards Premarital Sex` into the *x-axis label* box in the *Options* tab, replacing an automatically generated axis label, denoted by `<auto>`; because the axis label is longer than the text box, the scrollbar below the label is activated. Clicking *OK* opens a graphics-device window with the graph shown in Figure 3.13.

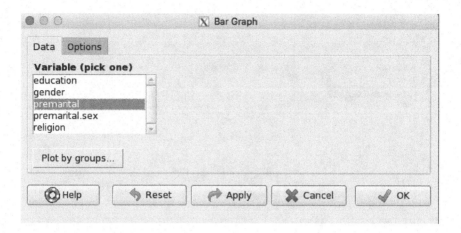

FIGURE 3.12: The *Bar Graph* dialog as it appears under Mac OS X (showing only the *Data* tab).

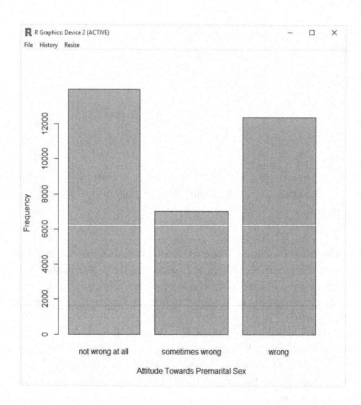

FIGURE 3.13: R graphics-device window with a bar graph for recoded attitude towards premarital sex.

3.5 Making a Contingency Table

I'm now ready to construct a contingency table to examine the relationship between the factors decade and premarital. From the R Commander menus, I choose *Statistics > Contingency tables > Two-way table*, producing the dialog box in Figure 3.14. In the *Data* tab, shown in the upper panel of the figure, I select the row and column variables for the table, premarital and decade, respectively. The *Subset expression* box near the bottom of the tab can be used to make a table for a subset of the full data set. For example, entering gender == "male" in this box would restrict the table to male respondents. Notice the double equals sign (==) for testing equality and the use of quotes around the factor level "male".[13]

The lower panel of Figure 3.14 shows the *Statistics* tab. Here, I press the radio button for *Column percentages* because decade, the column variable, is the explanatory ("independent") variable, and premarital, the row variable, is the response ("dependent") variable in the contingency table: It is standard practice to compute percentages within categories of the explanatory variable so as to make comparisons among these categories. The default in the dialog is *No percentages*. I leave the *Chi-square test* box checked—it's checked by default.[14] Clicking *OK* produces R commands and associated output in the R Commander *Output* pane; the commands and output are shown in Figure 3.15.

Examining the percentage table, it's evident that disapproval of premarital sex has declined over time, and, from the chi-square test, the relationship in the table is highly statistically significant: The p value for the chi-square test statistic in the printout, given as p-value < 2.2e-16, is to be read as $p < 2.2 \times 10^{-16}$, that is, less than 0.00000000000000022 (15 zeroes to the right of the decimal point followed by 22)—effectively 0. It is common for computer software like R to report very large or (as here) very small numbers in this format, called *scientific notation*.

In addition to the chi-square test and statistics associated with it (chi-square components and expected frequencies), the *Statistics* tab in the *Two-Way Table* dialog includes an option for computing Fisher's exact test for association in a contingency table.

---

[13]The double equals, ==, is used to test equality because the ordinary equals sign (=) is used for other purposes in R—in particular, to assign a value to a function argument (as in log(100, base=10)) or to assign a value to a variable (as in x = 10). Most R programmers prefer to use the left-pointing arrow (<-) for the latter purpose (as in x <- 10, read "the variable x *gets*—or is assigned—the value 10."). R expressions, including relational operators such as ==, are discussed in Section 4.4.2.

[14]If you're unfamiliar with the chi-square test of independence in a two-way table, don't worry; you'll almost surely study it in your introductory statistics course.

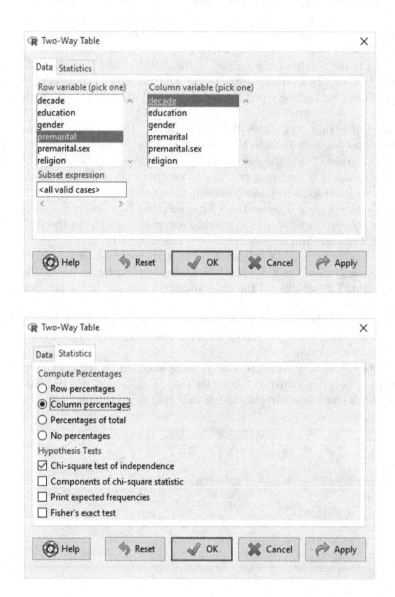

FIGURE 3.14: The *Two-Way Table* dialog, *Data* tab (top) and *Statistics* tab (bottom).

```
> local({
+    .Table <- xtabs(~premarital+decade, data=GSS)
+    cat("\nFrequency table:\n")
+    print(.Table)
+    cat("\nColumn percentages:\n")
+    print(colPercents(.Table))
+    .Test <- chisq.test(.Table, correct=FALSE)
+    print(.Test)
+ })

Frequency table:
                  decade
premarital          1970s 1980s 1990s 2000s
  not wrong at all   2423  3348  3647  4547
  sometimes wrong    1692  1789  1797  1738
  wrong              3207  3017  3035  3114

Column percentages:
                  decade
premarital          1970s   1980s   1990s   2000s
  not wrong at all   33.1    41.1    43.0    48.4
  sometimes wrong    23.1    21.9    21.2    18.5
  wrong              43.8    37.0    35.8    33.1
  Total             100.0   100.0   100.0   100.0
  Count            7322.0  8154.0  8479.0  9399.0

        Pearson's Chi-squared test

data:  .Table
X-squared = 413.3, df = 6, p-value < 2.2e-16
```

FIGURE 3.15: Contingency table for attitude towards premarital sex by decade.

3.6   Creating a Report

I've explained how R commands accumulate in the *R Script* tab as your R Commander session progresses. These commands can be edited and saved in a `.R` file via the R Commander menus, *File > Save script* or *File > Save script as*. As well, at the end of the session, the R Commander offers to save the script (see Section 3.8). A saved script can be reloaded into the *R Script* tab in a subsequent R Commander session, or used in an R editor independent of the R Commander, such as RStudio (discussed briefly in Section 1.4).

3.6.1   Creating a Report by Cutting and Pasting

The text in the *Output* pane is also editable, and you can copy and paste text from the *Ouput* pane into a text-editor or word-processor document to create a simple record of your work: Just use the R Commander *Edit* menu; the right-click context menu, after first clicking in the *Output* pane; or standard edit-key combinations.[15] If you go this route, however, be sure to use a monospaced (typewriter) font such as `Courier`, or your R output won't be properly aligned. The text in the *Output* pane can also be saved to a file, via *File > Save output* or *File > Save output as*.

You can similarly save graphs (such as the histogram in Figure 3.13 on page 34) from an R graphics device. Under Windows, graphs can be copied to the clipboard and subsequently pasted into a word-processor document or saved in a graphics file and subsequently imported into a document. To save a graph, use either the *File* menu in the graphics device or the right-click context menu. If you activate *History > Recording* from the Windows R graphics device, you'll be able to scroll through graphs in the graphics device via the *Page Up* and *Page Down* keys.

On Mac OS X, the R Commander creates graphs in a *Quartz* graphics-device window. The *Quartz* graphics device also supports copying to the clipboard via the key-combination *command-c*, and you can save graphs to PDF files via *File > Save As*. A graph copied to the clipboard can be pasted into most Mac word processors via *command-v*; similarly, a graph saved as a PDF file can typically be imported into a word processor document. Plot history is saved by default in the *Quartz* graphics device, and you can move back and forth among your graphs via the *command-←* and *command-→* key combinations.

3.6.2   Creating a Report as a Dynamic Document

In addition to the relatively crude approach of copying and pasting R output and graphs, the R Commander supports writing reports in the simple Markdown mark-up language.[16] Just as R commands accumulate during your session in the *R Script* tab, they are also written into the *R Markdown* tab. The advantage of using R Markdown in comparison to cutting and pasting output is that the R Markdown document that you create is a permanent, reproducible record of your work, intermixing executable R commands (essentially, the contents of the R script for your session) with your explanatory text. The resulting R Markdown document is then compiled into a report that includes R commands along with

---

[15]On Windows or Linux/Unix, you can use the key combinations *Ctrl-x* for cut, *Ctrl-c* for copy, and *Ctrl-v* for paste. On Mac OS X, you can use *command-x* for cut, *command-c* for copy, and *command-v* for paste, in addition to the various *control*-key combinations. See Section 3.7 for more information on editing in the R Commander.

[16]The R Commander also supports reports written in the more sophisticated LATEX mark-up language: See Section 3.9 on customizing the R Commander.

associated printouts and graphs. The contents of the *R Markdown* tab can also be saved (via *File* > *Save R Markdown file* or *File* > *Save R Markdown file as*), to be reloaded and reused in a subsequent R Commander session or in a compatible R editor, such as RStudio (see Section 1.4).

The *R Markdown* tab begins with one of two (customizable) R Markdown templates, depending upon whether you have installed the optional auxiliary Pandoc software:[17] If Pandoc is installed, the R Commander uses the newer **rmarkdown** package (Allaire et al., 2015a) to convert the R Markdown document into a Word file, an HTML file (a "web" page), or (if LATEX is additionally installed) a PDF file. If Pandoc isn't installed on your computer, the R Commander uses the older **markdown** package (Allaire et al., 2015b) to convert the R Markdown document into an HTML file. The initial contents of these alternative R Commander R Markdown templates are shown in Figure 3.16.

Both forms of the R Markdown template begin with title, author, and date fields. As likely is obvious, you should replace the generic title (`Replace with Main Title`) with your own descriptive title (as in the example below), and `Your Name` with your name. Leave the date field alone—in both templates a date and time stamp will be generated automatically— unless you want to hard-code the date. For the **rmarkdown** version of the template, you must retain the pairs of quotes (`"..."`) in the title, author, and date fields.

Both R Markdown templates then include a *block of commands* to customize the document and to load packages necessary for executing subsequent commands. This command block—which in both templates begins with a line of the form ```` ```{r `` *etc.*`}` and ends with the line ```` ``` ```` (i.e., three *back-ticks*)—should be left as is (unless you know what you're doing!).

With few exceptions (such as R commands that require direct user intervention), each time the R Commander generates a command or set of commands, they are entered into an R command block in the *R Markdown* tab, delimited by ```` ```{r} ```` at the top of the block and ```` ``` ```` at the bottom. Except in two cases, you probably shouldn't alter these command blocks (again, unless you know what you're doing):

- You should feel free to delete an *entire* command block, including the initial ```` ```{r} ```` and terminating ```` ``` ````, either by directly editing the text in the *R Markdown* tab or via the R Commander menus: *Edit* > *Remove last Markdown command block*, which deletes the command block generated by the preceding R Commander action. You may wish to do this, for example, if you generate incorrect or unwanted output. You should be careful, however, to insure that deleting a command block doesn't disturb the logic of the R session: For example, you shouldn't delete a block that reads a data set and retain subsequent blocks that perform computations on the data set.

- You can control the size of graphs drawn in a command block by adding `fig.height` and `fig.width` *arguments* to the initial ```` ```{r} ```` line. For example, ```` ```{r fig.height=4, fig.width=6} ```` sets figure height to 4 inches and width to 6 inches.

As in the *R Script* tab, you can edit text in the *R Markdown* tab by typing in the tab, by using the *Edit* menu when the cursor is in the tab, by right-clicking when the cursor is in the tab and selecting edit actions from the resulting context menu, or by using standard edit key combinations. Generally more conveniently, however, you can open an editor window via *Edit* > *Edit R Markdown document*, by right-clicking in the *R Markdown* tab when the cursor is in the tab and selecting *Edit R Markdown document* from the context menu, or by the key combination *Ctrl-e* when the cursor is in the *R Markdown* tab.[18] The R Markdown document editor for the current session is shown in Figure 3.17.

---

[17]See the instructions for installing optional software in Section 2.5.

[18]Under Mac OS X, you can also use *command-e*.

```
---
title: "Replace with Main Title"
author: "Your Name"
date: "AUTOMATIC"
---


```{r echo=FALSE, message=FALSE}
# include this code chunk as-is to set options
knitr::opts_chunk$set(comment=NA, prompt=TRUE)
library(Rcmdr)
library(car)
library(RcmdrMisc)
```
```

(a) Template for use with **rmarkdown** (and requiring Pandoc).

```
<!-- R Commander Markdown Template -->

Replace with Main Title
=======================

### Your Name

### `r as.character(Sys.Date())`

```{r echo=FALSE}
# include this code chunk as-is to set options
knitr::opts_chunk$set(comment=NA, prompt=TRUE,
          out.width=750, fig.height=8, fig.width=8)
library(Rcmdr)
library(car)
library(RcmdrMisc)
```
```

(b) Template for use with **markdown** (in the absence of Pandoc).

FIGURE 3.16: R Markdown templates used in the R Commander.

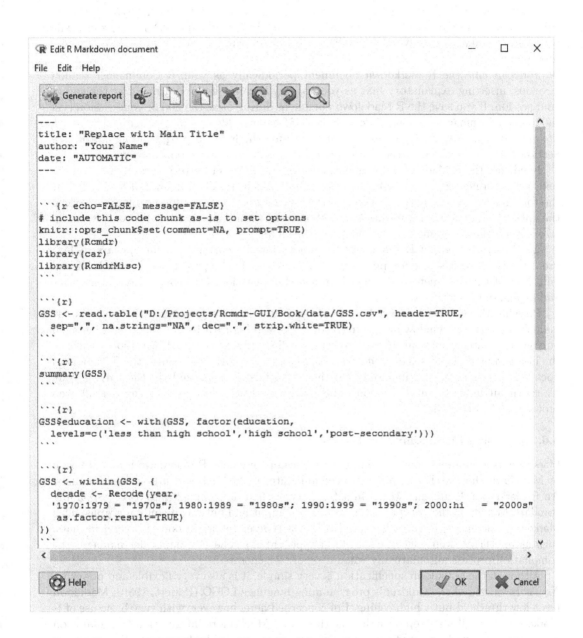

FIGURE 3.17: The R Markdown editor window.

The R Markdown editor includes *File*, *Edit*, and *Help* menus, the last providing help both on using R Markdown and on the editor itself, a toolbar with buttons for common editing operations (hover your mouse above the buttons to see "tool tips" describing the action associated with each button) along with a button for generating a report (described below), and *Help*, *OK*, and *Cancel* buttons at the bottom of the window. Clicking *OK* closes the editor, saving your edits, while clicking *Cancel* closes the editor and discards your edits. The editor is a "modal" dialog: While the editor window is open, operation of the R Commander is suspended.

You can edit the R Markdown document periodically as your R Commander session develops, inserting explanatory text as you go, or you can edit the document at the end of your session. If you save the R Markdown document during or at the end of your session (via the main R Commander menus, *File > Save R Markdown file* or *File > Save R Markdown file as*), you can edit it in any text editor, including the RStudio programming editor (see Section 1.4): The file is saved as a plain-text document with file extension .Rmd.

In editing the R Markdown document, however, you should be very careful only to type text *between* different R command blocks, each of which, recall, is delimited by ```{r} at the top and ``` at the bottom. Typing arbitrary text *within* a command block—between the initial ```{r} and terminating ``` of the block—almost inevitably will cause R syntax errors when the commands in the block are executed.

An illustrative edited R Markdown document for the current session appears in Figure 3.18. Because it's just for purposes of illustration, I've kept this document brief and only part of the document is displayed; in a real application, I'd include more descriptive and explanatory text.

Pressing the *Generate report* button in the editor or—if the editor isn't open—in the main *R Commander* window brings up the dialog box in Figure 3.19. If Pandoc isn't installed on your computer,[19] pressing *Generate report* will simply create an HTML report without the intervening dialog. I select *.html (web page)* and click *OK*. That causes the R Markdown document to be compiled, including running embedded R commands in the various code blocks in an independent R session. The resulting .html file opens in the default web browser, as in Figure 3.20.

3.6.2.1   Using Markdown: The Basics

Markdown is a punningly named, simple text markup language. R Markdown is an extension of Markdown that, as I've explained, accommodates embedded, executable R commands. An R Markdown document, stored in a file of type .Rmd, is compiled into a corresponding standard Markdown file of type .md, containing R input and output, including graphs. This Markdown document, in turn, is compiled into a typeset report in one of several formats, such as an HTML web page (i.e., a file of type .html). The R Commander manages the compilation process automatically when a report is generated.

Although the Markdown specification is very simple, it is also very flexible and powerful: As has been said of the children's programming language LOGO (Papert, 1980), Markdown has a low threshold but a high ceiling. I'm concerned here, however, with very basic use of R Markdown, so we'll just step over the low threshold. Much more information is available on line at http://rmarkdown.rstudio.com/, a web site that's accessible through the R Commander *Help* menu.

Basic Markdown syntax is illustrated in Figure 3.21, alongside the corresponding typeset HTML document. Some of this basic Markdown syntax is used in the R Markdown document for the current R Commander session (in Figure 3.18).

---

[19]See Section 2.5 for information on installing Pandoc.

```
---
title: "Contingency Table Example from Ch. 1"
author: "John Fox"
date: "AUTOMATIC"
---


```{r echo=FALSE, message=FALSE}
# include this code chunk as-is to set options
knitr::opts_chunk$set(comment=NA, prompt=TRUE)
library(Rcmdr)
library(car)
library(RcmdrMisc)
```


Reading and Summarizing the GSS data

```{r}
GSS <- read.table("D:/Projects/Rcmdr-GUI/Book/data/GSS.csv", header=TRUE,
  sep=",", na.strings="NA", dec=".", strip.white=TRUE)
```


```{r}
summary(GSS)
```


. . .

Constructing the Contingency Table

Cross-classifying attitude towards premarital sex by decade.
Because the explanatory variable is decade, the *column* variable,
I computed *column* percentages.

```{r}
local({
  .Table <- xtabs(~premarital+decade, data=GSS)
  cat("\nFrequency table:\n")
  print(.Table)
  cat("\nColumn percentages:\n")
  print(colPercents(.Table))
  .Test <- chisq.test(.Table, correct=FALSE)
  print(.Test)
})
```
As the decades progress, disapproval of premarital sex declines.
The relationship between the two variables is highly statistically
significant, with chi-square = 43.3, $df = 4$, $p < 2.2 \times 10^{-16}$.
```

FIGURE 3.18: An edited R Markdown document for the example in Chapter 3, showing only part of the document (with elided lines marked by . . .).

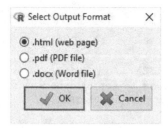

FIGURE 3.19: The *Select Output Format* dialog for creating a report from the R Commander R Markdown document.

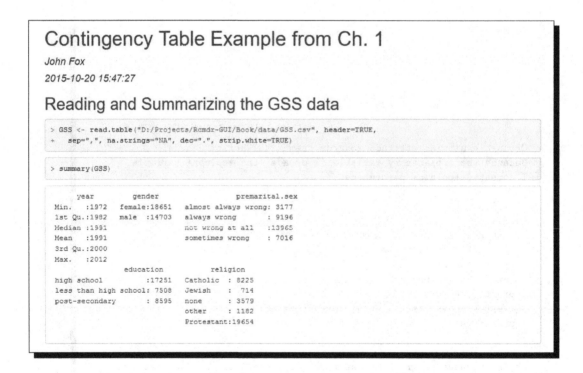

Contingency Table Example from Ch. 1

John Fox
2015-10-20 15:47:27

Reading and Summarizing the GSS data

```
> GSS <- read.table("D:/Projects/Rcmdr-GUI/Book/data/GSS.csv", header=TRUE,
+    sep=",", na.strings="NA", dec=".", strip.white=TRUE)
```

```
> summary(GSS)
```

```
      year          gender                  premarital.sex
 Min.   :1972   female:18651   almost always wrong: 3177
 1st Qu.:1982   male  :14703   always wrong       : 9196
 Median :1991                  not wrong at all   :13965
 Mean   :1991                  sometimes wrong    : 7016
 3rd Qu.:2000
 Max.   :2012
                education              religion
 high school          :17251   Catholic  : 8225
 less than high school: 7508   Jewish    :  714
 post-secondary       : 8595   none      : 3579
                              other      : 1182
                              Protestant:19654
```

FIGURE 3.20: The compiled HTML report, as it appears in a web browser, with only the top of the page shown.

```
---
title: "Markdown Illustrations"
---

Level 1 Heading
===================

Level 2 Heading
--------------

# Level 1 Heading

## Level 2 Heading

### Level 3 Heading

Paragraphs can be continued on
as many lines as desired.

Different paragraphs are separated by
a blank line of text.

Bullet items are preceded by *
and sub-items are preceded by
a tab and +:

* list item 1
    + sub-item 1.1
    + sub-item 1.2
* list item 2
    + sub-item 2.1
    + sub-item 2.2

Numbered lists are specified similarly
by starting each line with an
item number:

1. first item
    + sub-item 1.1
    + sub-item 1.2
2. second item
    + sub-item 2.1
    + sub-item 2.2

Text surrounded by * (*asterisks*) or
_ (_underscores_) is set in *italics*.

Text surrounded by ** (**double-asterisks**)
or __ (__double-underscores__) is set
in **boldface**.
```

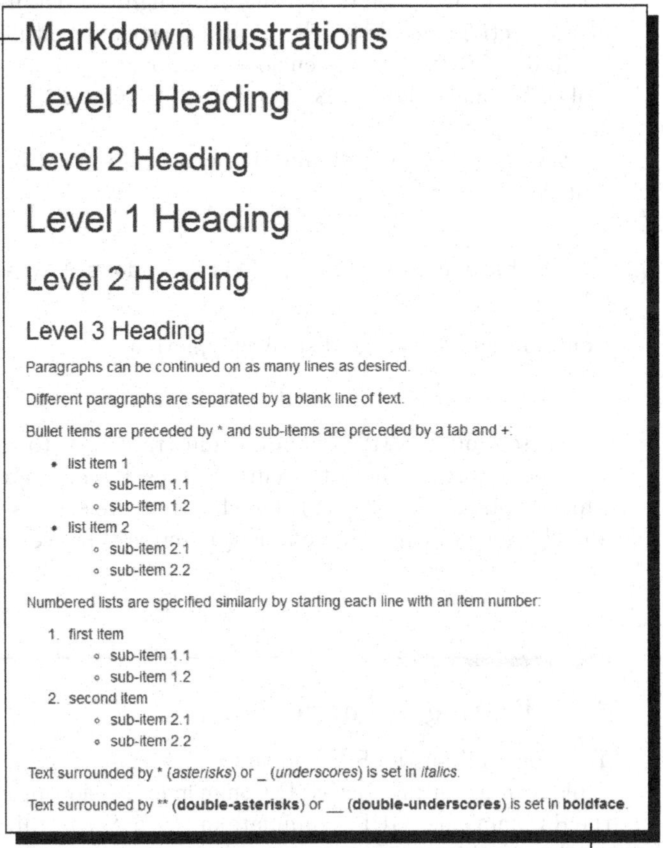

Markdown Illustrations

Level 1 Heading

Level 2 Heading

Level 1 Heading

Level 2 Heading

Level 3 Heading

Paragraphs can be continued on as many lines as desired.

Different paragraphs are separated by a blank line of text.

Bullet items are preceded by * and sub-items are preceded by a tab and +:

- list item 1
  - sub-item 1.1
  - sub-item 1.2
- list item 2
  - sub-item 2.1
  - sub-item 2.2

Numbered lists are specified similarly by starting each line with an item number:

1. first item
   - sub-item 1.1
   - sub-item 1.2
2. second item
   - sub-item 2.1
   - sub-item 2.2

Text surrounded by * (*asterisks*) or _ (*underscores*) is set in *italics*.

Text surrounded by ** (**double-asterisks**) or __ (**double-underscores**) is set in **boldface**.

FIGURE 3.21: Basic Markdown syntax. A Markdown file is shown on the left and the corresponding HTML page on the right.

3.6.2.2   Adding a Little LaTeX Math to an R Markdown Document*

The last line of the document in Figure 3.18 also demonstrates how to imbed LaTeX math inside an R Markdown document. Although the details of LaTeX are well beyond the scope of this section, simple LaTeX math is reasonably intuitive.[20]

In-line LaTeX math is enclosed in dollar-signs ($...$). In the example document, the embedded math `$df = 4$`, `$p < 2.2 \times 10^{-16}$` is typeset as $df = 4$, $p < 2.2 \times 10^{-16}$.

Similarly, a displayed equation can be specified using double dollar signs ($$). For example,

```
$$
y_i = \beta_0 + \beta_1 x_{1i} + \cdots + \beta_k x_{ki} + \epsilon_i
$$
```

would be typeset as the displayed equation

$$y_i = \beta_0 + \beta_1 x_{1i} + \cdots + \beta_k x_{ki} + \epsilon_i$$

These simple LaTeX examples illustrate how to use underscores for subscripts, as in `y_i` and `x_{1i}` (where the curly braces { and } are used for grouping), and carets for superscripts, as in `10^{-16}`. Greek letters are specified as `\beta` (β), `\epsilon` (ϵ), and so on. The LaTeX symbol `\cdots` is typeset as three centered dots ($\cdots$).

3.7   Editing Commands*

You can think of the R Commander *R Script* tab as a simple programming editor. As I've explained, as an interactive R Commander session progresses, the commands generated by the R Commander GUI accumulate in the *R Script* tab. The resulting R script of commands can be saved in a .R file, to be reloaded into the R Commander in a subsequent session—or into another R programming editor, such as RStudio (discussed briefly in Section 1.4)—to be modified or re-executed.

You can also type R commands directly into the R Commander *Script* tab, or modify commands previously generated by the GUI. The *R Script* tab isn't a full-featured programming editor, but it does support basic editing functions, such as *cut, copy, paste, undo*, and so on, via a largely self-explanatory right-click context menu (shown at the left of Figure 3.22), the R Commander *Edit* menu (see Figure A.1 on page 200), and standard key-combinations (with the available key bindings listed in Table 3.2).

To provide a simple example of R-command editing, I return to the bar graph constructed for attitude towards premarital sex in the GSS data set, displayed in Figure 3.13 (on page 34). This graph was created by the R command

```
with(GSS, Barplot(premarital, xlab="Attitude Towards Premarital Sex",
  ylab="Frequency"))
```

which, along with the other commands generated during the current R Commander session, appears in the *R Script* tab. As a general matter, computing in R is performed by *functions*, which are called by name with their *arguments* in parentheses. The arguments may be given by position or may be named, with each named argument associated with a value by an equals sign (=). In this example, two functions are called: `with` and `Barplot`.

---

[20]If you're interested in pursuing the topic, a good place to start is the *Wikipedia* article on LaTeX, at https://en.wikipedia.org/wiki/LaTeX, which includes several useful links and references.

 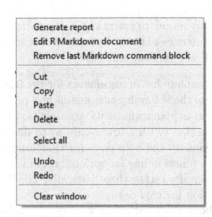

FIGURE 3.22: *R Script* tab right-click context menu (left) and *R Markdown* tab right-click context menu (right).

TABLE 3.2: R Commander edit-key bindings.

Key-Combination	Action
Ctrl-x	cut selection to clipboard
Ctrl-c	copy selection to clipboard
Ctrl-v	paste selection from clipboard
Ctrl-z or *Alt-backspace*	undo last operation (may be repeated)
Ctrl-w	redo last undo
Ctrl-f or *F3*	open find-text dialog
Ctrl-a	select all text
Ctrl-s	save file
Ctrl-r or *Ctrl-Tab*	submit ("run") current line or selected lines (*R Script* tab)
Ctrl-e	open document editor (*R Markdown* tab)

Except as noted, these key bindings work in the *R Script* and *R Markdown* tabs and in the *Output* and *Messages* panes. Under Mac OS X, either the *command* or *control* key may be used. On keyboards with "function" keys, the *F3* function key may be used as an alternative to *Ctrl-f*.

- The `with` function takes two arguments, and both are specified here by position: The first argument is a data set, `GSS` in the example. The second argument is an *expression* referencing variables in the data set—in this case, a call to the `Barplot` function.

- The `Barplot` function is called with three arguments, the first specified by position and the other two by name: `premarital` is the variable in the `GSS` data set to used for the bar graph; the arguments `xlab` and `ylab` are character strings (enclosed in quotes) specifying the horizontal and vertical axis labels.

To see the complete list of arguments for the `Barplot` function, once again select *Graphs > Bar graph* from the R Commander menus and press the *Help* button in the resulting dialog. In addition to an explanation of its arguments, you'll see that the `Barplot` function calls the `barplot` function (with a lower-case b) to draw the graph.[21] Clicking on the hyperlink for `barplot` brings up the help page for that function.

The `barplot` function has an optional argument, `horiz`, which, if set to `TRUE`, draws the bar graph horizontally rather than vertically. The R Commander *Bar Graph* dialog, however, makes no provision for this option.

To draw a horizontal bar graph of attitude towards premarital sex, I left-click and drag the mouse over the original `Barplot` command in the *R Script* tab, selecting the command, copy this text by *Ctrl-c*, and paste it by *Ctrl-v* at the bottom of the script.[22] I then edit the command to draw the horizontal bar graph:

```
with(GSS, Barplot(premarital, ylab="Attitude Towards Premarital Sex",
  xlab="Frequency", horiz=TRUE))
box()
```

In addition to setting `horiz=TRUE`, I also exchange the `xlab` and `ylab` arguments, so that the axes are properly labelled, and I add a second command, a call to the `box` function (with no arguments[23]), to draw a box around the plotting region in the graph. Selecting these commands with the mouse, I press the *Submit* button, obtaining the modified, horizontal bar graph in Figure 3.23.

It's a limitation of the R Commander *R Script* tab that you have to submit *complete* commands: You may continue a command over as many lines as necessary, and you may simultaneously submit more than one complete command (as I've done in this example), but submitting a partially complete command results in an error. Moreover, submitting a command or commands for execution in this manner from the *R Script* tab also causes the commands to be entered into the document in the *R Markdown* tab.

R command blocks automatically incorporated into the *R Markdown* tab (discussed in Section 3.6) are similarly modifiable. Moreover, if you know what you're doing, you can write your own R command blocks. It's important to pay attention to the order of commands, however: For example, you can't use a data set in a computation before you input the data.

---

[21] The `Barplot` function resides in the **RcmdrMisc** package, which is loaded when the R Commander starts, while the `barplot` function is in the **graphics** package, which is a standard part of R.

[22] Alternatively, I could have edited the original command in place and resubmitted it for execution, but by copying the command I retain both versions.

[23] Even though the `box()` command has no arguments, it's still necessary to include the parentheses so R knows that this is a function call.

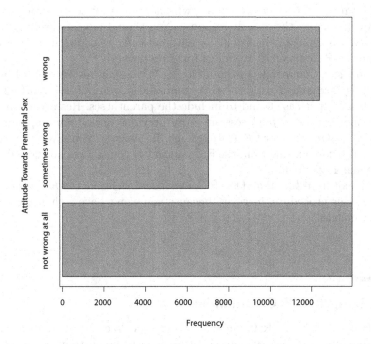

FIGURE 3.23: Horizontal bar graph of attitude towards premarital sex in the GSS data set. This graph was created by modifying the Barplot command produced by the R Commander *Bar Graph* dialog.

3.8   Terminating the **R Commander** Session

In most cases, the simplest and safest method of terminating an R Commander session is to select *File > Exit > From Commander and R* from the R Commander menus. In addition to closing the *R Commander* window, you will also end your R session without saving the R workspace. On exit, the R Commander will give you an opportunity to save the contents of the *R Script* tab, the *R Markdown* tab, and the *Output* pane, however.

You can also exit from the R Commander *without* terminating your R session by selecting *File > Exit > From Commander*. Again, the R Commander will prompt you to save your work. You can subsequently exit from the *R Console* via *File > Exit* on Windows or *File > Close* on Mac OS X.

Upon exiting from the *R Console*, R will ask whether you want to save your workspace, with saving the workspace as the default response. You should almost surely *not* save the workspace. A saved workspace, which will be automatically reloaded in a subsequent R session, may cause the R Commander to fail to function properly.[24]

If you exit from the R Commander without closing R, you can restart the R Commander interface by entering `Commander()` at the R > command prompt. It's important to spell `Commander()` with an upper-case C and to include the parentheses. Restarting the R Commander in this manner initiates a *fresh* session—your previous work doesn't appear in the *R Script* tab, the *R Markdown* tab, or the *Output* pane. If, however, you saved the script or R Markdown document before exiting from the R Commander, you can reload these documents via the R Commander *File* menu.

Finally, you can exit from *both* R and the R Commander by closing the *R Console* directly. I don't recommend this procedure, however, because you won't have a chance to save your work in the R Commander.

3.9   Customizing the **R Commander***

The default configuration of the R Commander should be fine for most users, but some aspects of the appearance and behavior of the software can be customized to reflect your preferences and needs. Specific R Commander features are set via the R `options` command, and can be saved so that they are persistent across R Commander sessions.

3.9.1   Using the *Commander Options* Dialog

The most convenient way to set many R Commander options is via *Tools > Options*, which produces the dialog box in Figure 3.24. The several tabs in this dialog show the default selections, some of which vary from one operating system to another: For example, the default *Theme* in the *Other Options* tab is *vista* on Windows and *clearlooks* on Mac OS X.

Most of the settings in the *Commander Options* dialog are self-explanatory, and you can experiment with the settings to see their effects on the R Commander. A few comments on the options are in order, however:

---

[24]If you experience problems caused by an inadvertently saved workspace, you can remove the `.RData` file containing the saved workspace by following the troubleshooting instructions in Section 2.2.1 for Windows, 2.3.4 for Mac OS X, or 2.4.1 for Linux/Unix.

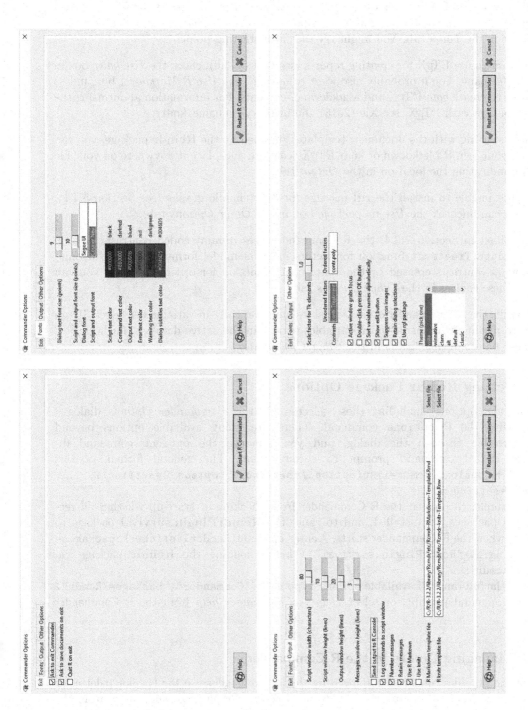

FIGURE 3.24: The *Commander Options* dialog, with *Exit*, *Fonts*, *Output*, and *Other Options* tabs.

- Clicking on one of the text color-selection buttons in the *Fonts* tab brings up a color-selection sub-dialog. See the discussion of colors in Section 3.9.3.

- To display the R Commander on a screen for a presentation with a data projector, I typically set the *Dialog text font size* to 14 points and the *Script and output font size* to 15 points in the *Fonts* tab. You might try that as a starting point.

- If you want to use LaTeX for creating reports (see Section 3.6), check the *Use knitr* box in the *Output* pane. You'll probably also want to *uncheck* the *Use R Markdown* box, unless you wish to create *both* LaTeX and Markdown documents. For information about using the **knitr** package with LaTeX, see Xie (2015) and http://yihui.name/knitr/.

- Possibly starting with the document templates supplied by the **Rcmdr** package, you can prepare your own R Markdown or knitr LaTeX template and place it anywhere on your file system, indicating the location in the *Output* tab.

- If you are unable to install the **rgl** package for 3D dynamic graphs (see Section 5.4.1), then you can uncheck the *Use rgl package* box in the *Other Options* tab.

- As described in Section 7.2.4, the R Commander uses dummy-coded contrasts created by the `contr.Treatment` function for factors in linear-model formulas and orthogonal-polynomial contrasts created by the `contr.poly` function for ordered factors. You can change these choices in the *Other Options* tab.

- By default, the R Commander orders variables alphabetically in variable list boxes. If you prefer to retain the order in which variables appear in the active data set, uncheck *Sort variables names alphabetically* in the *Other Options* tab.

3.9.2   Setting Rcmdr Package Options

R Commander options—including those selected via the *Commander Options* dialog—are set with the R `options` command. There are many available options beyond those accessible through the dialog, and you can use the `options` command directly at the R command prompt to specify them. The general format of this command is `options(Rcmdr=list(`*option.1=setting.1, option.2=setting.2, ..., option.n=setting.n*`))`.

For example, to prevent the R Commander from checking at start-up whether all recommended packages are installed, and to cause the **RcmdrPlugin.survival** package to be loaded when the R Commander starts,[25] enter `options(Rcmdr=list(check.packages=FALSE, plugins="RcmdrPlugin.survival"))` before loading the **Rcmdr** package via `library(Rcmdr)`.

To see the full range of available options, type `help("Commander", package="Rcmdr")` at the R command prompt, or select *Help > Commander help* from the R Commander menus.

3.9.3   Managing Colors in the R Commander

Many of the graphics functions employed by the R Commander use the R color palette for color selection. You can modify the palette by selecting *Graphs > Color palette* from the R Commander menus, producing the dialog box at the top of Figure 3.25, which initially displays the current palette. Typically, the first color in the palette (*black* in the default

---

[25]See Chapter 9 for a discussion of R Commander plug-in packages.

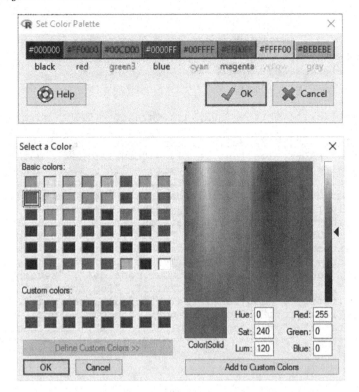

FIGURE 3.25: *Set Color Palette* dialog (top) and the Windows *Select a Color* sub-dialog (bottom). A color version of this figure appears in the insert at the center of the book.

palette) is used for most graphical elements, with the remaining colors used successively. You may wish to change the default colors if you are red-green color blind, for example.

Pressing one of the color buttons—for example, the second (*red*) button—brings up the *Select a Color* sub-dialog shown at the bottom of Figure 3.25. The structure of the color-selection dialog varies by operating system but its use is simple and direct: In the Windows version of the dialog, displayed in the figure, you can select a basic color by left-clicking on it, select a color by clicking in the color-selection box or moving the slider at the right, define a color by hue, saturation, and luminosity, or define a color by the intensity (0–255) of its red, green, and blue additive primary-color components.

R maintains a list of more than 650 named colors. If you choose a color that's near a named color, then the name appears underneath the color button in the *Set Color Palette* dialog. As you can see, all eight of the colors in the default palette have names.

3.9.4   Saving R Commander Options

R has an elaborate start-up process employing several configuration files.[26] Users who just want to customize the R Commander, however, can cut through the details: If you make selections in the *Commander Options* dialog or employ the R `options` command directly to configure the R Commander, then simply select *Tools > Save Rcmdr options* from the R Commander menus, producing the dialog in Figure 3.26. Prior to this menu selection, I

---

[26]Enter the command ?`Startup` at the command prompt in the R console for an explanation of R's start-up procedure.

FIGURE 3.26: The R Commander *Save Commander Options* editor. The dialog edits or creates the file `.Profile` in your home directory.

changed the R Commander dialog text font size to 14 points, and the script and output font size to 15 points.

The *Save Commander Options* dialog in Figure 3.26 is a simple text editor. The dialog creates the R configuration file `.Rprofile` in your home directory, where it will be found when you next start R. If you *already* have a `.Rprofile` file in your home directory, then the dialog will modify it with your current R Commander options—those lines in the file between

`###! Rcmdr Options Begin !###`

and

`###! Rcmdr Options End !###`

Near the bottom of this block of lines is a command that you can "uncomment" (by removing the #s) to start the R Commander whenever R starts up:

```
# Uncomment the following 4 lines (remove the #s)
#   to start the R Commander automatically when R starts:

# local({
#     old <- getOption('defaultPackages')
#     options(defaultPackages = c(old, 'Rcmdr'))
# })
```

Any other contents of a pre-existing `.Rprofile` file are undisturbed.

4

Data Input and Data Management

This chapter shows how to get data into the R Commander from a variety of sources, including entering data directly at the keyboard, reading data from a plain-text file, accessing data stored in an R package, and importing data from an Excel or other spreadsheet, or from other statistical software. I also explain how to save and export R data sets from the R Commander, and how to modify data—for example, how to create new variables and how to subset the active data set.

4.1 Introduction to Data Management

Managing data is an admittedly dry but nevertheless vitally important topic. As experienced researchers will attest, collecting, assembling, and preparing data for analysis are typically much more time consuming than analyzing the data. It makes sense to take up data management early because you'll always have to read data, and usually have to modify them, before you can perform statistical analysis. I suggest that you read as much of this chapter as you need to get started with your work, but that you at least familiarize yourself with the balance of its contents, so that you know where to look when a particular data management problem presents itself.

A graphical interface such as the R Commander is frankly not the best data management tool. Even common data management tasks are very diverse, and data sets often have unique characteristics that require customized treatment. GUIs, in contrast, excel at tasks where the choices are limited and can be anticipated.

As a flexible programming language, R is well suited to data management tasks, and there are many R packages that can help you with these tasks. Often, the most straightforward solution is to write a script of R commands or a simple R program to prepare a data set for analysis. The data management script or program need not be elegant or efficient; it simply must work properly, because it typically will be used only once. What does it matter if your data management script takes a few minutes to run? That's just a small part of the time you'll invest in preparing and analyzing your data. Learning a little bit of R programming (see Section 1.4), therefore, goes a long way to making you a more efficient data analyst, even if you choose to use the R Commander for routine data analysis tasks.

4.2 Data Input

Recall from Chapter 3 that data sets in the R Commander are stored in R data frames—rectangular data sets in which the rows represent cases and the columns are usually either numeric variables or factors (categorical variables). There are many ways to read data into the R Commander, and I will describe several of them in the following subsections, including entering data directly at the keyboard into the R Commander data editor, reading data from plain-text files, importing data from other software including spreadsheets such as Excel, and accessing data sets stored in R packages.

Plain-text or spreadsheet data must be in a simple, rectangular form, with one row or line for each case, and one column or "field" for each variable. There may be an initial row with variable names, or an initial column with case names, or both. If the data are in an irregular or other more complex form, you'll likely have to do some work before you can read them into the R Commander. The R Commander, however, can handle simple merges of separate data sets with the same variables or cases (see Section 4.5).

4.2.1   Entering Data into the R Commander Data Editor

One way to enter a small data set into the R Commander is to type the data into a plain-text file using an editor or into a spreadsheet, and then to use the input methods described in Section 4.2.2 or 4.2.3. My personal preference is to keep small data sets in plain-text files.

Alternatively, it is also simple to type small data sets directly into the R Commander data editor. To illustrate, I'll enter a data set that appears in an exercise in an introductory statistics text, Moore et al. (2013, Exercise 5.4).[1] The data, for 12 women in a study of dieting, comprise two numeric variables: each woman's lean body mass, in kilograms, and her resting metabolic rate, in calories burned per hour.

I begin by selecting *Data > New data set* from the R Commander menus.[2] That brings up the *New Data Set* dialog shown at the top-left of Figure 4.1. I replace the default data-set name Dataset with the more descriptive name Metabolism. Pressing *OK* opens the R Commander data editor with an empty data set, as shown at the top-right. There is initially one case (row) and one variable (column) in the data set.

Pressing the *Add column* button once and the *Add row* button 11 times produces the still-empty data set shown at the bottom-left of Figure 4.1, with all of the values in the body of the data table initially missing (NA). You can also at any point press the *Enter* or *return* key to add a new row to the bottom of the data set, or the *Tab* key to add a new column at the right of the data. Using these keys rather than the buttons can be convenient for initial data entry: Simply *Tab* as you enter numbers in the first row until the data set contains columns for all the variables, and then *Enter* (or *return*) after you complete typing in the data in each row.

I next replace the generic variable names V1 and V2 with mass and rate, and enter the values for each variable, as given in the text by Moore et al., into the cells of the data table, as shown at the bottom-right. Once finished, I press the *OK* button in the editor, making Metabolism the active data set. The following message appears in the R Commander *Messages* pane: NOTE: The dataset Metabolism has 12 rows and 2 columns.

To see whether and how the variables are related, I suggest that you make a scatterplot of the data, via *Graphs > Scatterplot*, with lean body mass on the horizontal axis and metabolic rate on the vertical axis. If you wish, then perform a linear least-squares regression of rate on mass: *Statistics > Fit models > Linear regression.*[3]

---

[1]This is a newer edition of the basic statistics text that I used as the original "target" for the R Commander, in that I aimed to cover all of the methods described in the text. As I mentioned in Chapter 1, however, the R Commander has since broadened its scope.

[2]The R Commander session in each chapter of this book is independent of the sessions in preceding chapters; if you're following along with the book on your computer, restart R and the R Commander for each chapter.

[3]See Section 5.4.1 for more information about making scatterplots in the R Commander, and Section 7.1 on least-squares regression.

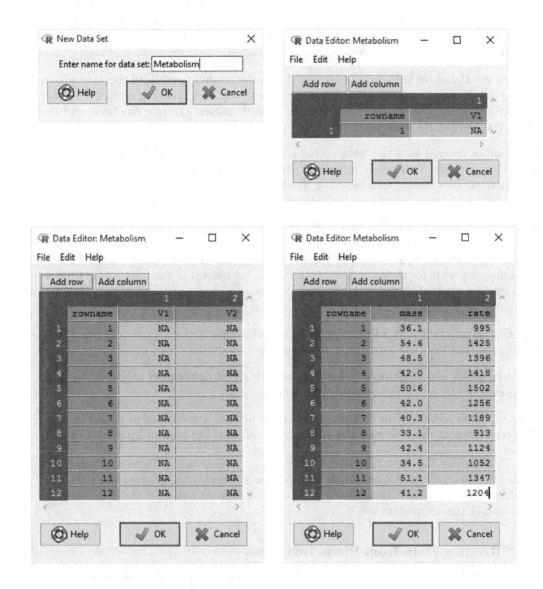

FIGURE 4.1: Using the R Commander data editor to enter the Metabolism data set, proceeding from upper-left to lower-right.

Here is some additional information about using the R Commander data editor for direct data entry:

- In this data set, the row names are simply numbers supplied by the editor, but you can also type names into the `rownames` column of the data editor. If you do so, however, make sure that each name is unique (i.e., that there are no duplicate names) and that any names with embedded blanks are enclosed in quotes (e.g., `"John Smith"`); alternatively, just squeeze out the blanks (`JohnSmith`), or use a period, dash, or other character as a separator (`John.Smith`, `John-Smith`).

- You can use the arrow keys on your keyboard to move around the cells of the data editor, or simply left-click in any cell. When you do so, text that you type will replace whatever is currently in the cell—initially, the default variable names, the row numbers, and the `NA`s in the body of the data table.

- If a column entered into the data editor consists entirely of numbers, it will become a numeric variable. Conversely, a column containing *any* non-numeric data (with the exception of the missing data indicator `NA`) will become a factor. Character data with embedded blanks must be enclosed in quotes (e.g., `"agree strongly"`).

- If the initial column width of the data editor cells is insufficient to contain a variable name, a row name, or a data value, the cell in question won't be displayed properly. Left-click on either the left or right border of the data editor and drag the border until all cells are sufficiently wide for their contents. If the width (or height) of the data editor window is insufficient to display simultaneously all columns (or rows) of the data set, then the horizontal (or vertical) scrollbar will be activated.

- The *Edit* menu in the data editor supports some additional actions, such as deleting rows or columns, and the *Help* menu provides access to information about using the editor.

- The data editor is a *modal dialog*: Interaction with the R Commander is suspended until you press the *OK* or *Cancel* button in the editor.

- After typing in a data set in this manner, it is, as always, a good idea to press the *View data set* button in the R Commander toolbar to confirm that the data have been entered correctly.

You can also use the R Commander data editor to modify an *existing* data set—for example, to fix an incorrect data value: Press the *Edit data set* button in the R Commander toolbar to edit the active data set.

4.2.2 Reading Data from Plain-Text Files

In Section 3.3, I demonstrated how to read data into the R Commander from a plain-text, comma-separated-values (*CSV*) data file. To recapitulate, each line in a CSV file represents one case in the data set, and all lines have the same number of values, separated by commas. The first value in each line may be a case name, and the first line of the data file may contain variable names, also separated by commas. If there are *both* case names and variable names in the data, then there will be one fewer variable name in the first line than values in subsequent lines; otherwise, there must be the same number of *fields* (values) in each line of the file. On input to R, empty fields (i.e., produced by adjacent commas) or fields consisting only of spaces will translate into missing data (`NA`s).

CSV files are a kind of least common denominator of data storage. Almost all software that deals with rectangular data sets, including spreadsheet programs like Excel and statistical software like SPSS, are capable of reading and writing CSV files. CSV files, therefore, are often the simplest file format for moving data from one program to another.

You may have to do minor editing of a CSV file produced by another program before reading the data into the R Commander—for example, you may find it convenient to change missing data codes to NA[4]—but these operations are generally straightforward in any text editor. Do use an editor intended for plain-text (ASCII) files, however:[5] Word processors, such as Word, store documents in special files that include formatting information, and these files cannot in general be read as plain text. If you must edit a data file in a word processor, be very careful to save the file as plain text.

An advantage of CSV files is that character data fields may include embedded blanks (as, e.g., in `strongly agree`) without having to enclose the fields in quotes (`"strongly agree"` or `'strongly agree'`). Fields that include commas, however, must be quoted (e.g., `"don't know, refused"`). By the way, fields that include single (`'`) or double (`"`) quotes must be quoted using the *other* quote character (as in `"don't know, refused"`); otherwise, either single or double quotes may be used (as long as the *same* quote character is used at both ends of a character value—e.g., `"strongly agree'` isn't legal).

The R Commander *Read Text Data* dialog also supports reading data from plain-text files with fields separated by white space (one or more blanks), tabs, or arbitrary characters, such as colons (`:`) or semicolons (`;`).[6] Moreover, the data file may reside on the Internet rather than the user's computer, or may be copied to and read from the clipboard (as illustrated for spreadsheet data in Section 4.2.3).

Figure 4.2 shows a few lines from a small illustrative text data file, `Duncan.txt`, with white space separating data fields. Multiple spaces are employed to make the data values line up vertically, but this is inessential—one space would suffice to separate adjacent data values. Periods are used in the case names instead of spaces (e.g., `mail.carrier`) so that the names need not be quoted; commas (`,`) and dashes (`-`) are similarly used for the categories of `type`.[7]

The data, on 45 U.S. occupations in 1950, are drawn mostly from Duncan (1961); I added the variable type of occupation to the data set. The variables are defined in Table 4.1. The income and education data were derived by Duncan from the U.S. Census, while occupational prestige was obtained from ratings of the occupations in a social survey of the population. Duncan used the least-squares linear regression of `prestige` on `income` and `education` to compute predicted prestige scores for the majority of occupations in the Census for which there weren't direct prestige ratings.[8]

---

[4]The *Read Text Data* dialog allows you to specify an input missing data indicator different from NA (which is the default), but it will not accommodate *different* missing data codes for different variables or *multiple* missing data codes for an individual variable. In these cases, you could redefine other codes as missing data after reading the data, using *Data > Manage variables in active data set > Recode variables* (see Sections 3.4 and 4.4.1), or, as suggested, edit the data file prior to reading it into the R Commander to change all missing data codes to NA or another common value.

[5]There are many plain-text editors available. Windows systems come with the Notepad editor, and Mac OS X with TextEdit. If you enter data in TextEdit on Mac OS X, be sure to convert the data file to plain text, via *Format > Make Plain Text*, prior to saving it. The RStudio programming editor for R (discussed in Section 1.4) can also be used to edit plain-text data files.

[6]Both CSV files and data files that employ other field delimiters are plain-text files. Conventionally, the file type (or extension) `.csv` is used for comma-separated data files, while the file type `.txt` is used for other plain-text data files.

[7]`Duncan.txt` and other files used in this chapter may be downloaded from the web site for the text, as described in the Section 1.5.

[8]Duncan's occupational prestige regression is partly of interest because it represents a relatively early use of least-squares regression in sociology, and because Duncan's methodology is still employed for construct-

```
                      type            income education prestige
   accountant         prof,tech,manag    62      86       82
   pilot              prof,tech,manag    72      76       83
   architect          prof,tech,manag    75      92       90
   . . .
   bookkeeper         white-collar       29      72       39
   mail.carrier       white-collar       48      55       34
   insurance.agent    white-collar       55      71       41
   . . .
   janitor            blue-collar         7      20        8
   policeman          blue-collar        34      47       41
   waiter             blue-collar         8      32       10
```

FIGURE 4.2: The `Duncan.txt` file, with white-space-delimited data from Duncan (1961) on 45 U.S. occupations in 1950. Only a few of the 46 lines in the file are shown; the widely spaced ellipses (. . .) represent elided lines. The first line in the file contains variable names. I added the variable `type` to the data set.

TABLE 4.1: Variables in Duncan's occupational prestige data set.

Variable	Values
`type`	`blue-collar`; `white-collar`; `prof,tech,manag` (professional, technical, or managerial)
`income`	percentage of occupational incumbents earning $3500 or more
`education`	percentage of occupational incumbents with high-school education or more
`prestige`	percentage of prestige ratings of good or better

TABLE 4.2: Variables in the Canadian occupational prestige data set.

Variable	Values
education	average years of education of occupational incumbents
income	average annual income of occupational incumbents, in dollars
prestige	average prestige rating of the occupation (0–100 scale)
women	percentage of occupational incumbents who were women
census	the Census occupation code
type	bc, blue-collar; wc, white-collar; prof, professional, technical, or managerial

Reading a white-space-delimited, plain-text data file in the R Commander is almost identical to reading a comma-delimited file: Select *Data > Import data > from text file, clipboard, or URL* from the R Commander menus. Fill in the resulting dialog box (see Figure 3.4 on page 25) to reflect the structure and location of the input file. In the case of Duncan.txt, I would take all of the defaults, including the default *White space* field separator, with the exception of the name of the data set, where I'd substitute a descriptive name like Duncan for the default name Dataset.

4.2.3 Importing Data from Spreadsheets and Other Sources

Many researchers enter, store, and share small data sets in spreadsheet files. To illustrate, I've prepared two Excel files, the older format file Datasets.xls and the newer format Datasets.xlsx, both containing two data sets: Duncan's U.S. occupational prestige data, and a similar data set for Canada circa 1970, which is described by Fox and Suschnigg (1989). The Canadian occupational prestige data include the variables in Table 4.2, and the spreadsheet containing the data appears in Figure 4.3, which shows the first 21 rows in the spreadsheet; there are 103 rows in all, including the initial row of variable names. The education, income, and occupational gender composition data come from the 1971 Canadian Census, while the prestige scores are the average ratings of the occupations on a 0–100 "thermometer" scale in a mid-1960s Canadian national survey. The structure of the spreadsheet is similar to that of a plain-text input file: There is one row in the spreadsheet for each case, there's an optional row of variable names at the top, and there's an optional initial column of case names. When case names are present in the first column, there should be no variable name at the top of the column.

To read the Excel spreadsheet, I select *Data > Import data > from Excel file* from the R Commander menus, which brings up the dialog box at the left of Figure 4.4. I complete the dialog box to reflect the structure of the *Prestige* spreadsheet—including retaining the default *<empty cell> Missing data indicator*—and I enter the descriptive name Prestige for the data set, replacing the generic default name Dataset.[9]

Pressing *OK* in the *Import Excel Data Set* dialog leads to a standard *Open* file dialog box, where I navigate to the location of the Excel file containing the data, select it, and press the *Open* button, producing the *Select one table* dialog at the right of Figure 4.4. I left-click

---

ing socioeconomic-status scales for occupations. For further discussion of this regression, see Fox (2016, especially Chapter 11).

[9]Four of the occupations (athletes, newsboys, babysitters, and farmers) have missing occupational type, and the corresponding cells of the spreadsheet are empty. If, for example, NA were used to represent missing data in the spreadsheet, I'd type NA as the *Missing data indicator* in the dialog box.

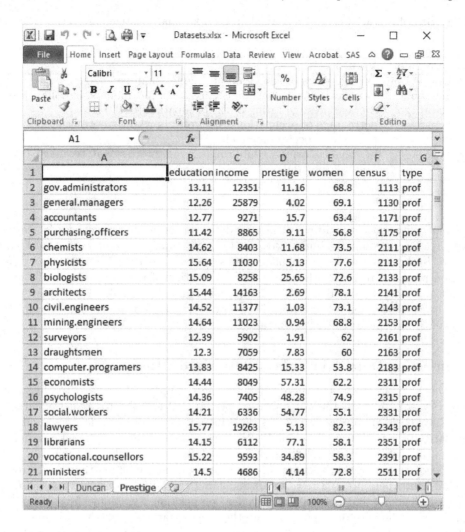

FIGURE 4.3: The Excel file `Datasets.xlsx` showing the first 21 (of 103) rows in the *Prestige* spreadsheet.

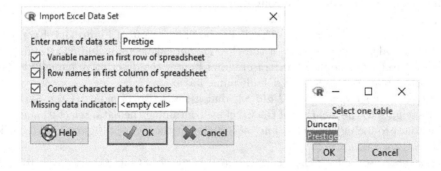

FIGURE 4.4: The *Import Excel Data Set* dialog box (left), and the *Select one table* sub-dialog (right), choosing the *Prestige* spreadsheet (table).

on the *Prestige* table and click *OK* to read the data into the R Commander, making the resulting Prestige data frame the active data set.

There are two other simple procedures for reading rectangular data stored in spreadsheets:

- Export the spreadsheet as a comma-delimited plain-text file. Editing the spreadsheet before exporting it, to make sure that the data are in a suitable form for input to the R Commander, can save some work later. For example, you don't want commas within cells unless the contents of the cells are quoted.

  Likewise, you might have to edit the resulting CSV file before importing the data into the R Commander. For example, if the first row of the spreadsheet contains variable names and the first column contains row names, the exported CSV file will have an empty first field corresponding to the empty cell at the upper-left of the spreadsheet (as in the *Prestige* spreadsheet in Figure 4.3). Simply delete the extra initial comma in the first row of the CSV file; otherwise, the first column will be treated as a variable rather than as row names.

- Alternatively, select the cells in the spreadsheet that you want to import into the R Commander: You can do this by left-clicking and dragging your mouse over the cells, or by left-clicking in the upper-left cell of the selection and then *Shift*-left-clicking in the lower-right cell. Selecting cells in this manner is illustrated in Figure 4.5 for the Excel spreadsheet containing Duncan's occupational prestige data. Then copy the selection to the clipboard in the usual manner (e.g., *Ctrl-c* on a Windows system or *command-c* on a Mac); in the R Commander, choose *Data > Import data > from text file, clipboard, or URL,* and press the *Clipboard* radio button in the resulting dialog, leaving the default *White space* selected as the *Field Separator*. The data are read from the clipboard as if they reside in a white-space-separated plain-text file.

In addition to plain-text files and Excel spreadsheets, the *Data > Import data* menu includes menu items to read data from SPSS internal and portable files, from SAS xport files, from Minitab data files, and from Stata data files. You can practice with the SPSS portable file Nations.por, which is on the web site for this book.

4.2.4 Accessing Data Sets in R Packages

Many R packages include data sets, usually in the form of R data frames, and these are suitable for use in the R Commander. When the R Commander starts up, some packages containing data sets are loaded by default. If you've installed other packages with data that you want to use, you can load the packages subsequently via *Tools > Load package(s)*.

Selecting *Data > Data in packages > List data sets in packages* from the R Commander menus opens a window listing all data sets available in currently loaded R packages. Selecting *Data > Data in packages > Read data from an attached package* produces the dialog box in Figure 4.6. Initially, the dialog appears as at the top of the figure.

If you know the name of the data set you want to read, you can type it into the *Enter name of data set* box, as in the middle of Figure 4.6, where I entered the name of the Duncan data set. As it turns out, this data set, containing Duncan's occupational prestige data, is supplied by the **car** package (Fox and Weisberg, 2011).[10] Because data sets in R packages are associated with documentation, pressing the *Help on selected data set* button will now bring up the help page for the Duncan data set. Clicking *OK* reads the data and makes Duncan the active data set in the R Commander.

---

[10]In subsequent chapters, I'll frequently draw data sets from the **car** package for examples.

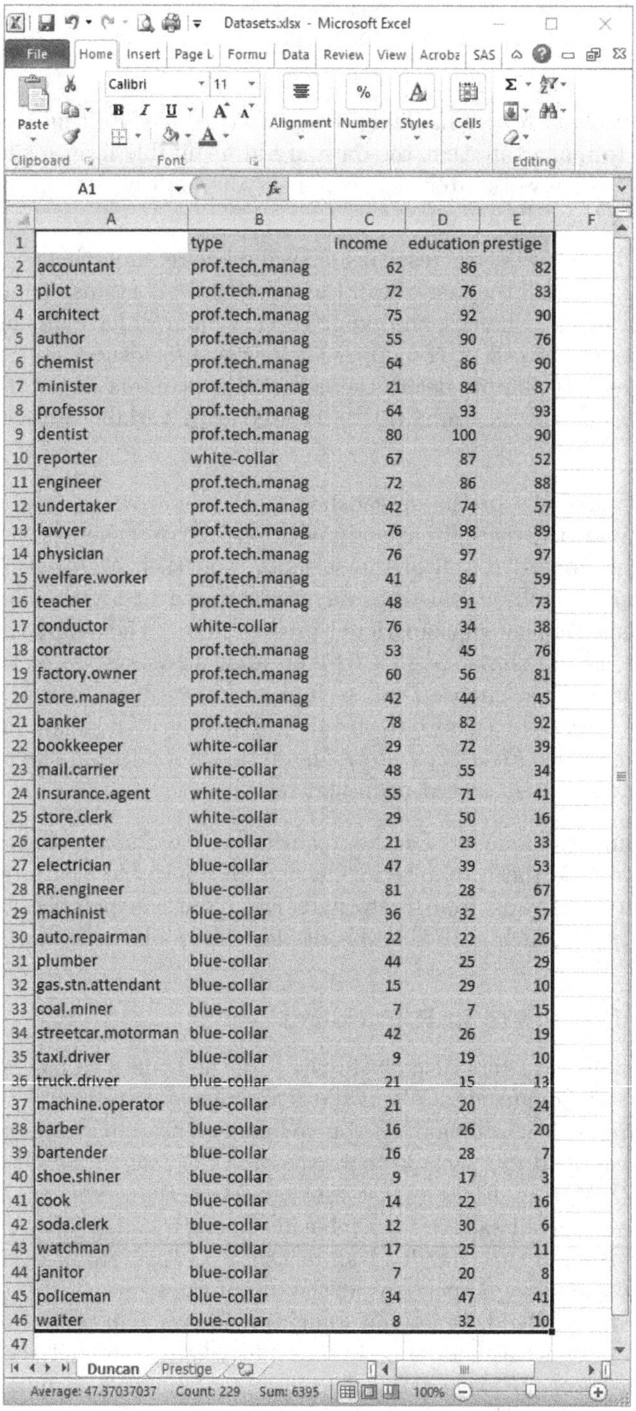

Datasets.xlsx - Microsoft Excel

File | Home | Insert | Page L | Formu | Data | Review | View | Acroba | SAS

A1

	A	B	C	D	E	F
1		type	income	education	prestige	
2	accountant	prof.tech.manag	62	86	82	
3	pilot	prof.tech.manag	72	76	83	
4	architect	prof.tech.manag	75	92	90	
5	author	prof.tech.manag	55	90	76	
6	chemist	prof.tech.manag	64	86	90	
7	minister	prof.tech.manag	21	84	87	
8	professor	prof.tech.manag	64	93	93	
9	dentist	prof.tech.manag	80	100	90	
10	reporter	white-collar	67	87	52	
11	engineer	prof.tech.manag	72	86	88	
12	undertaker	prof.tech.manag	42	74	57	
13	lawyer	prof.tech.manag	76	98	89	
14	physician	prof.tech.manag	76	97	97	
15	welfare.worker	prof.tech.manag	41	84	59	
16	teacher	prof.tech.manag	48	91	73	
17	conductor	white-collar	76	34	38	
18	contractor	prof.tech.manag	53	45	76	
19	factory.owner	prof.tech.manag	60	56	81	
20	store.manager	prof.tech.manag	42	44	45	
21	banker	prof.tech.manag	78	82	92	
22	bookkeeper	white-collar	29	72	39	
23	mail.carrier	white-collar	48	55	34	
24	insurance.agent	white-collar	55	71	41	
25	store.clerk	white-collar	29	50	16	
26	carpenter	blue-collar	21	23	33	
27	electrician	blue-collar	47	39	53	
28	RR.engineer	blue-collar	81	28	67	
29	machinist	blue-collar	36	32	57	
30	auto.repairman	blue-collar	22	22	26	
31	plumber	blue-collar	44	25	29	
32	gas.stn.attendant	blue-collar	15	29	10	
33	coal.miner	blue-collar	7	7	15	
34	streetcar.motorman	blue-collar	42	26	19	
35	taxi.driver	blue-collar	9	19	10	
36	truck.driver	blue-collar	21	15	13	
37	machine.operator	blue-collar	21	20	24	
38	barber	blue-collar	16	26	20	
39	bartender	blue-collar	16	28	7	
40	shoe.shiner	blue-collar	9	17	3	
41	cook	blue-collar	14	22	16	
42	soda.clerk	blue-collar	12	30	6	
43	watchman	blue-collar	17	25	11	
44	janitor	blue-collar	7	20	8	
45	policeman	blue-collar	34	47	41	
46	waiter	blue-collar	8	32	10	
47						

◄ ◄ ► ►| **Duncan**   Prestige

Average: 47.37037037   Count: 229   Sum: 6395   100%

FIGURE 4.5: Selecting the cells in Duncan's data set from the *Duncan* spreadsheet, to be copied to the clipboard.

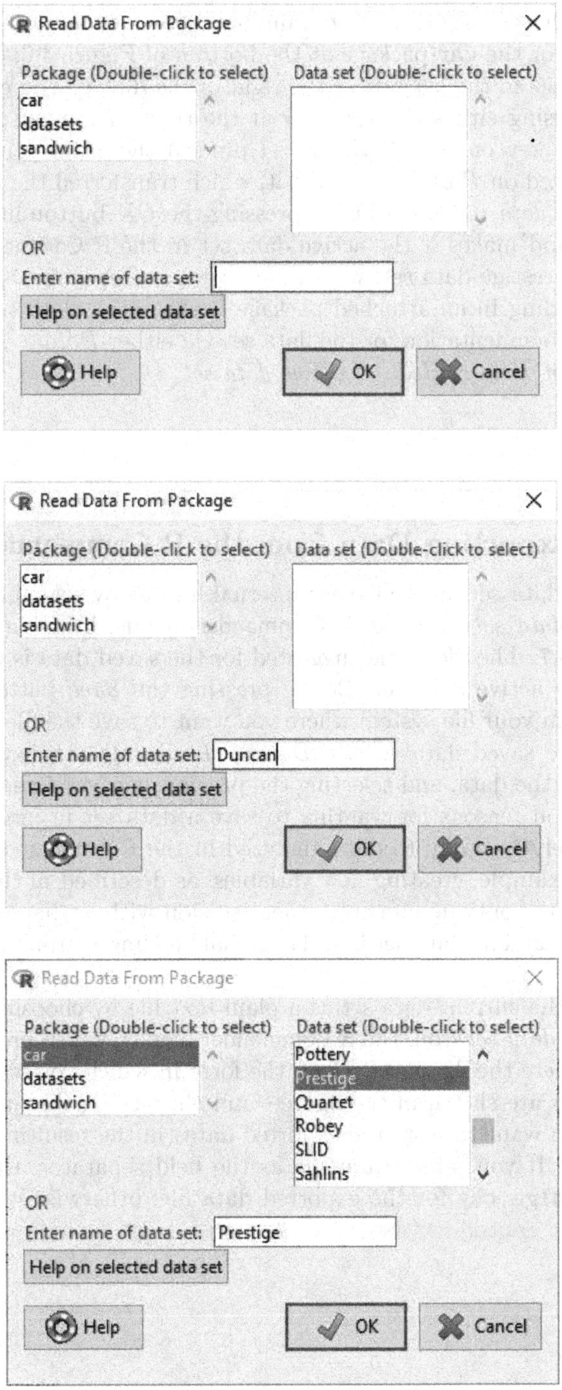

FIGURE 4.6: Reading a data set from an R package. The initial state of the *Read Data From Package* dialog is shown at the top; in the middle, I typed `Duncan` as the data set name; at the bottom, I selected *car* from the package list and *Prestige* from the data set list.

Alternatively, the left-hand list box in the *Read Data From Package* dialog includes the names of currently attached packages that contain data sets. Double-clicking on a package in this list, as I've done for the **car** package at the bottom of Figure 4.6, displays the data sets in the selected package in the list box at the right of the dialog. You can scroll this list in the normal manner, using either the scrollbar at the right of the list or clicking in the list and pressing a letter key on your keyboard—I pressed the letter *p* in Figure 4.6 and subsequently double-clicked on *Prestige* to select it, which transferred the name of the data set to the *Enter name of data set* box. Finally, pressing the *OK* button in the dialog reads the `Prestige` data set and makes it the active data set in the R Commander. This is the Canadian occupational prestige data set, with which we are already familiar.

When a data set residing in an attached package is the active data set in the R Commander, you can access documentation for the data set via either *Data > Active data set > Help on active data set* or *Help > Help on active data set*.

4.3   Saving and Exporting Data from the **R Commander**

You can save the active data set in an efficient internal format by selecting *Data > Active data set > Save active data set* from the R Commander menus, bringing up the *Save As* dialog shown in Figure 4.7. The file name suggested for the saved data is `Prestige.RData` because `Prestige` is the active data set. Before pressing the *Save* button in the dialog, navigate to the location in your file system where you want to save the file. In a subsequent session, you can load the saved data set via *Data > Load data set*, navigating your file system to the location of the data, and selecting the previously saved `Prestige.Rdata` file.

There are two common reasons for wanting to save a data set in internal format, the second of which is unlikely to apply to data analyzed in the R Commander: (1) You have modified the data—for example, creating new variables, as described in the next section— and you want to be able to continue in a subsequent session without having to repeat your data management work. (2) The data set is so large that reading it from a plain-text file is time consuming.

You can also export the current data set as a plain-text file by choosing *Data > Active data set > Export active data set* from the R Commander menus, which produces the dialog box in Figure 4.8. Complete the dialog to reflect the form in which you want to export the data—the default choices are shown in the figure—and click *OK*, subsequently navigating to the location where you want to store the exported data (in the resulting *Save As* dialog box, which isn't shown). If you select *Commas* as the field separator, the R Commander suggests the name `Prestige.csv` for the exported data file; otherwise it suggests the file name `Prestige.txt`.

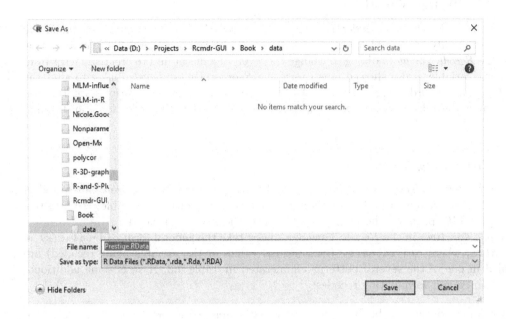

FIGURE 4.7: Saving the active data set.

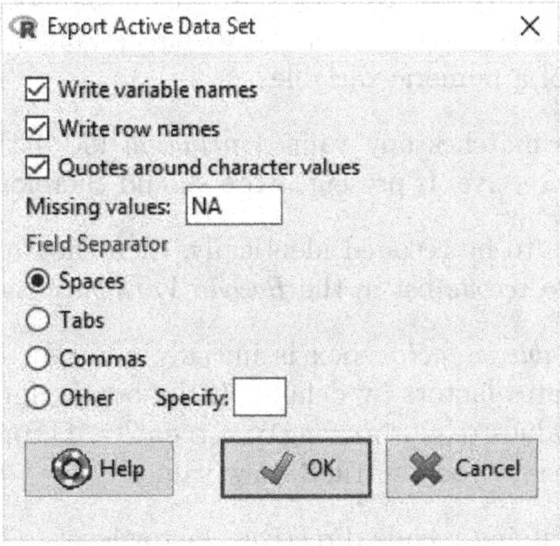

FIGURE 4.8: Exporting the active data set as a plain-text file.

4.4   Modifying Variables

Menu items in the R Commander *Data > Manage variables in active data set* menu are devoted to modifying variables and creating new variables. In Section 3.4, I explained how to use the *Recode Variables* dialog to change the levels of a factor and to create a factor from a numeric variable. I also showed how to use the *Reorder Factor Levels* dialog to alter the default alphabetic ordering of factor levels. In this section, I provide additional information on recoding variables and describe other facilities in the R Commander for transforming data.

4.4.1   Recoding Variables

Two common uses of the *Recode Variables* dialog to create new factors were illustrated in Section 3.4: Figure 3.9 (page 31) shows how to transform a numeric variable into a factor, and Figure 3.10 (page 32) shows how to reorganize the levels of a factor.

Recode directives in the *Recode Variables* dialog take the general form `old-value(s) = new-value`, where `old-value(s)` (i.e., original values of the variable being recoded) are specified in one of the several patterns enumerated in Table 4.3. Here is some additional information about formulating recodes:

- If an old value of the variable being recoded satisfies *none* of the recode directives, the value is simply carried over into the recoded variable. For example, if the directive `"strongly agree" = "agree"` is employed, but the old value `"agree"` is not recoded, then *both* old values `"strongly agree"` and `"agree"` map into the new value `"agree"`.

- If an old value of the variable to be recoded satisfies *more than one* of the recode directives, then the first applicable directive is applied. For example, if the variable `income` is recoded with `lo:25000 = "low"` and `25000:75000 = "middle"` (specified in that order), then a case with `income = 25000` receives the new value `"low"`.

- As illustrated in the recode directive `lo:25000 = "low"`, the special value `lo` may be used to represent the smallest value of a numeric variable; similarly, `hi` may be used to represent the largest value of a numeric variable.

- The special old value `else` matches any value (*including* NA[11]) that doesn't explicitly match a preceding recode directive. If present, `else` should therefore appear last.

- If there are several variables to be recoded identically, then they may be selected simultaneously in the *Variables to recode* list in the *Recode Variables* dialog.

- The *Make (each) new variable a factor* box is initially checked, and consequently the *Recode Variables* dialog creates factors by default. If the box is unchecked, however, you may specify numeric new values (e.g., `"strongly agree" = 1`), or even character (e.g., `1:10 = "low"`) or logical (e.g., `"yes" = TRUE`) new values.

- As a general matter, in specifying recode directives, factor levels and character values on both sides of = must be quoted using double quotes (e.g., `"low"`), while numeric values (e.g., `10`), logical values (`TRUE` or `FALSE`), and the missing data value `NA` are not quoted.

---

[11]Thus, if you want to retain NAs where they currently appear and still want to use `else`, then you can specify the directive `NA = NA` prior to `else`.

TABLE 4.3: Recode directives employed in the *Recode Variables* dialog.

Old Value(s)		Example Recode Directives
an individual value	*a*	`99 = NA`
		`NA = "missing"`
		`"strongly agree" = "agree"`
a set of values	*a, b, ..., k*	`1, 3, 5 = "odd"`
		`"strongly agree", "agree somewhat" = "agree"`
a numeric range	*a:b*	`1901:2000 = "20th Century"`
		`lo:20000 = "low income"`
		`100000:hi = "high income"`
anything else (must appear last)	`else`	`else = "other"`

The special old values `lo` and `hi` may be used to represent the smallest and largest values of a numeric variable, respectively.

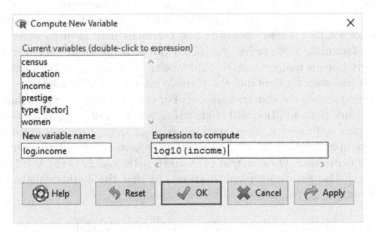

FIGURE 4.9: The *Compute New Variable* dialog.

- Recall that one recode directive is entered on each line of the *Recode directives* box: After you finish typing a recode directive, press the *Enter* or *Return* key to move to the next line.

4.4.2 Computing New Variables

Selecting *Data > Manage variables in active data set > Compute new variable* from the R Commander menus produces the dialog box displayed in Figure 4.9. At the top of the dialog is a list of variables in the active data set (the **Prestige** data set, read from the **car** package in Section 4.2.4). Notice that the variable **type** is identified as a factor in the variable list; the other variables in the data set are numeric.

At the bottom of the dialog are two text fields, the first of which contains the name of the new variable to be created (initially, **variable**): I typed **log.income** as the name of the new variable.[12] If the name of the new variable is the same as that of an existing variable in the current data set, then, when you press the *OK* or *Apply* button in the dialog, the R Commander will ask whether you want to replace the existing variable with the new one.

The second text box, which is initially empty, contains an R *expression* defining the new variable; you can double-click on variable names in the *Current variables* list to enter names into the expression, or you can simply type in the complete expression. In the example, I typed `log10(income)` to compute the log base 10 of `income`.[13] Pressing *OK* or *Apply* causes the expression to be evaluated, and (if there are no errors) the new variable `log.income` is added to the `Prestige` data set.

New variables may be simple transformations of existing variables (as in the log transformation of `income`), or they may be constructed straightforwardly from two or more existing variables. For example, the data set `DavisThin` in the **car** package contains seven variables, named DT1 to DT7, that compose a "drive-for-thinness" scale.[14] The items are each scored 0, 1, 2, or 3. To compute the scale, I need to sum the items, which I may do with the simple expression `DT1 + DT2 + DT3 + DT4 + DT5 + DT6 + DT7` (after reading the `DavisThin` data set into the R Commander, of course, as described in Section 4.2.4).

Table 4.4 displays R arithmetic, comparison, and logical operators, along with some commonly employed arithmetic functions. Both relational and logical operators return logical values (`TRUE` or `FALSE`). These operators and functions may be used individually or in combination to formulate more or less complex expressions to create new variables; for example, to convert Celsius temperature to Fahrenheit, `32 + 9*celsius/5` (supposing, of course, that the active data set contains the variable `celsius`). The precedence of operators in complex expressions is the conventional one: For example, multiplication and division have higher precedence than addition and subtraction, so `1 + 2*6` is 13. Where operators, such as multiplication and division, have equal precedence, an expression is evaluated from left to right; for example, `2/4*5` is 2.5. Parentheses can to used, if desired, to alter the order of evaluation of an expression: For example `(1 + 2)*6` is 18, and `2/(4*5)` is 0.1. When in doubt, parenthesize! Also remember (from Section 3.5) that the double equals sign (`==`), not the ordinary equals sign (`=`), is used to test for equality.

4.4.2.1 Complicated Expressions in Computing New Variables*

The *Compute New Variable* dialog is more powerful than it appears at first sight, because *any* R expression may be specified as long as it produces a variable with the same number of values as there are rows in the current data set. Suppose, for example, that the active data set is the `Prestige` data set, which includes the numeric variable `education`, in years. The following expression uses the `factor` and `ifelse` functions to recode `education` into a factor (as an alternative to the *Recode Variables* dialog):

```
factor(ifelse(education > 12, "post-secondary", "less than post-secondary"))
```

Here's another example of the use of `ifelse`, which selects the larger of husband's and wife's income in an imagined data set of heterosexual married couples:

```
ifelse(hincome > wincome, hincome, wincome)
```

---

[12]Remember the rules for naming variables in R: Names may only contain lower- and upper-case letters, numerals, periods, and underscores, and must begin with a letter or period.

[13]Don't be concerned if you're unfamiliar with logarithms (logs)—I'm using them here simply to illustrate a data transformation. Logs are often employed in data analysis to make the distribution of a strictly positive, positively skewed variable, such as income, more symmetric.

[14]These data were generously made available by Caroline Davis of York University in Toronto, who studies eating disorders.

TABLE 4.4: R operators and common functions useful for formulating expressions to compute new variables.

Symbol	Explanation	Examples
	Arithmetic Operators (return numbers)	
–	negation (unary minus)	`-loss`
+	addition	`husband.income + wife.income`
–	subtraction	`profit - loss`
*	multiplication	`hours.worked*wage.rate`
/	division	`population/area`
^	exponentiation	`age^2`
	Relational Operators (return TRUE or FALSE)	
<	less than	`age < 21`
<=	less than or equal to	`age <= 20`
==	equal to	`age == 21`
		`gender == "male"`
>=	greater than or equal to	`age >= 21`
>	greater than	`age > 20`
!=	not equal to	`age != 21`
		`marital.status != "married"`
	Logical Operators (return TRUE or FALSE)	
&	and	`age > 20 & gender == "male"`
|	or (inclusive)	`age < 21 | age > 65`
!	not (unary)	`!(age < 21 | age > 65)`
	Common Arithmetic Functions (return numbers)	
`log`	natural log	`log(income)`
`log10`	log base 10	`log10(income)`
`log2`	log base 2	`log2(income)`
`sqrt`	square root	`sqrt(area)` (equivalent to `area^0.5`)
`exp`	exponential function, e^x	`exp(rate)`
`round`	rounding	`round(income)` (to the nearest integer)
		`round(income, 2)` (to two decimal places)

If `hincome` exceeds `wincome`, then the corresponding value of `hincome` is used; otherwise, the corresponding value of `wincome` is returned.

The general form of the `ifelse` command is `ifelse(logical-expression, value-if-true, value-if-false)`, where

- *logical-expression* is a logical expression that evaluates to TRUE or FALSE for each case in the data set. Thus, in the first example above, `education > 12` evaluates to TRUE for those with more than 12 years of education and to FALSE for those with 12 or fewer years of education. In the second example, `hincome > wincome` evaluates to TRUE for couples for whom the husband's income exceeds the wife's income, and FALSE otherwise.

- *value-if-true* gives the value(s) to be assigned to cases for which *logical-expression* is TRUE. This may be a single value, as in the first example (the character string `"post-secondary"`), or a *vector* of values, one for each case, as in the second example (the vector of husbands' incomes for the couples); if *value-if-true* is a vector, then the *corresponding* entry of the vector is used where *logical-expression* is TRUE.

- *value-if-false* gives the value(s) to be assigned to cases for which *logical-expression* is FALSE; it too may be a single value (e.g., `"less than post-secondary"`), or a vector of values (e.g., `wincome`).

4.4.3 Other Operations on Variables

Most of the remaining items in the *Data > Manage variables in active data set* menu (see Figure A.3 on page 201) are reasonably straightforward:

- *Add observation numbers to data set* creates a new numeric variable named `ObsNumber`, with values $1, 2, \ldots, n$, where n is the number of rows in the active data set.

- *Standardize variables* transforms one or more numeric variables to mean 0 and standard deviation 1.

- *Reorder factor levels* permits you to change the default alphabetic ordering of factor levels, and was illustrated in Section 3.4: See in particular Figure 3.8 (page 30).

- It sometimes occurs—for example, after subsetting a data set (an operation described in Section 4.5)—that not all levels of a factor actually appear in the data. *Drop unused factor levels* removes empty levels, which occasionally cause problems in analyzing the data.

- *Rename variables* and *Delete variables from data set* do what they say.

- I discuss *Define contrasts for a factor* in Chapter 7 on statistical models in the R Commander (see Section 7.2.4).

The two remaining items in the *Manage variables in active data set* menu convert numeric variables to factors:

- *Bin numeric variable* allows you to categorize a possibly continuous numeric variable into class intervals, called *bins*. The resulting *Bin a Numeric Variable* dialog is shown in Figure 4.10, where I select `income` as the variable to bin; named the factor to be created `income.level` (the default name is `variable`), select 4 bins (the default is 3), opt for *Equal-count bins* (the default is *Equal-width bins*), and select *Numbers* for the level names (the default is to specify the level names in a sub-dialog). Clicking *OK* adds the factor `income.level` to the data set, where level `"1"` represents the (rough) fourth of cases with the lowest income, `"2"` the next fourth, and so on.

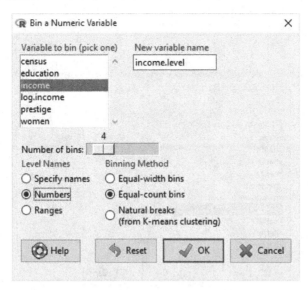

FIGURE 4.10: The *Bin a Numeric Variable* dialog, creating the factor `income.level` from the numeric variable `income`.

- Some data sets use numeric codes, typically consecutive integers (e.g., 1, 2, 3, etc.) to represent the values of categorical variables. Such variables will be treated as numeric when the data are read into the R Commander. *Convert numeric variables to factors* allows you to change these variables into factors, either using the numeric codes as level names ("1", "2", "3", etc.) or supplying level names directly (e.g., "strongly disagree", "disagree", "neutral", etc.).

I'll illustrate with the `UScereal` data set in the **MASS** package; this data set contains information about 65 brands of breakfast cereal marketed in the U.S.[15] To access the data set, I first load the **MASS** package by the menu selection *Tools > Load package(s)*, selecting the **MASS** package in the resulting dialog (Figure 4.11). I then input the `UScereal` data and make them the active data set via *Data > Data in packages > Read data set from an attached package* (as described in Section 4.2.4).

Figure 4.12 demonstrates the conversion of the numeric variable `shelf`—the supermarket shelf on which the cereal is displayed—originally coded 1, 2, or 3, to a factor with corresponding levels "low", "middle", and "high". Clicking *OK* in the main dialog (at the left of Figure 4.12) brings up the sub-dialog (at the right), into which I type the level names corresponding to the original numbers. Having been converted into a factor, the variable `shelf` can now be used, for example, in a contingency table in the R Commander (as described in Section 3.5), and will be treated appropriately as a categorical variable if it appears as a predictor in a regression model (see Sections 7.2.3 and 7.2.4).

---

[15]I'm grateful to an anonymous referee for suggesting this example.

FIGURE 4.11: The *Load Packages* dialog, selecting the **MASS** package.

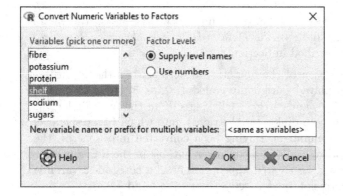 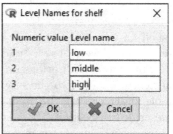

FIGURE 4.12: The *Convert Numeric Variables to Factors* dialog (left) and *Level names* sub-dialog (right), converting `shelf` in the `UScereal` data set to a factor.

4.5 Manipulating Data Sets

In contrast to the operations on variables discussed in the preceding section, selections in the *Data > Active data set* menu (see Figure A.3 on page 201) act on data sets as a whole or on rows of data sets. Some of the items in the *Active data set* menu are entirely straightforward, and I'll simply explain briefly what they do:

- *Select active data set* allows you to choose from among the data frames in your workspace, if there are more than one; selecting this menu item is equivalent to pressing the *Data set* button in the R Commander toolbar.

- *Refresh active data set* resets the information that the R Commander maintains about the active data set, such as the variable names in the data set, which variables are numeric and which are factors, and so on. This information is used, for example, in variable list boxes and to determine which menu items are active. You may need to refresh the active data set if you make a change to the data set *outside of* the R Commander menus—for example, if you type in and execute an R command that adds a variable to the data set. In contrast, when changes to the active data set are made via the R Commander GUI, the data set is refreshed automatically.

- *Help on active data set* opens the documentation for the data set if it was read from an R package.

- *Variables in active data set* lists the names of the variables in the data set in the *Output* pane.

- *Set case names* opens a dialog to set the row (case) names of the active data set to the values of a variable in the data set. This operation can be useful if the row names weren't established when the data were read into the R Commander. The row names variable may be a factor, a character variable, or a numeric variable, but its values must be unique (i.e., no two rows can have the same name). Once row names are assigned, the row names variable is removed from the data set.

- The actions performed by *Save active data set* and *Export active data set* were described in Section 4.3.

4.5.1 Special Operations on Data Sets*

Two of the items in the *Active data set* menu have specialized functions, and so I'll describe them briefly as well:

- *Aggregate variables in active data set* summarizes the values of one or more variables according to the levels of a factor, producing a new data set with one case for each level. Aggregation proceeds by applying some function—the mean, the sum, or another function that returns a single value—to the values of the variable for cases in each level of the factor. For example, starting with a data set in which the cases represent individual Canadians and that contains a factor for their province of residence, along with other variables such as years of education and dollars of annual income, you can produce a new data set in which the cases represent provinces and the variables include mean education and mean income in each province.

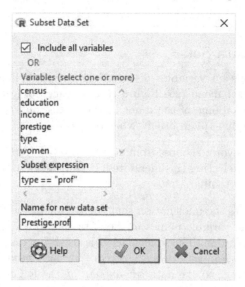

FIGURE 4.13: The *Subset Data Set* dialog.

- *Stack variables in active data set* creates a data set in which two or more variables are "stacked," one on top of the other, to produce a single variable. If there are n cases in the active data set, and if k variables are to be stacked, the new data set will contain one variable and $n \times k$ cases, along with a factor whose levels are the names of the stacked variables in the original data set. This transformation of the data set is occasionally useful in drawing graphs.

4.5.2 Subsetting Cases

Three items in the *Active data set* menu create subsets of cases, most directly *Subset active data set*, which brings up the dialog box shown in Figure 4.13, where the active data set is the Canadian occupational prestige data (`Prestige`) from the **car** package.[16] I complete the dialog so that the subsetted data will include all variables in the original data set, which is the default. I change the *Subset expression* from the default `<all cases>` to the logical expression `type == "prof"` to select professional, technical, and managerial occupations.[17] More generally, the subset expression should return a logical value for each case (`TRUE` or `FALSE`—see the discussion of R expressions in Section 4.4.2). I also change the default name of the new data set `<same as active data set>` to `Prestige.prof`.

The *Subset Data Set* dialog can also be used to create a subset of *variables* in the active data set: Just uncheck the *Include all variables* box, use the *Variables* list box to select the variables to be retained, and leave the *Subset expression* at the default `<all cases>`.

Remove row(s) from the active data set in the *Data > Active data set* menu leads to the dialog box in Figure 4.14, where the active data set is the `Duncan` occupational prestige data set from the **car** package. I type in the case names `"minister"` `"conductor"` and replace the default name for the new data set, `<same as active data set>`, with `Duncan.1`. Clicking `OK` deletes these two cases from the `Duncan` data, so `Duncan.1` contains

---

[16]As I've explained, you can make `Prestige` the active data set by pressing the *Data set* button in the R Commander toolbar, selecting `Prestige` from the list of data sets currently in memory.

[17]Recall that the double equals sign `==` is used in R to test equality.

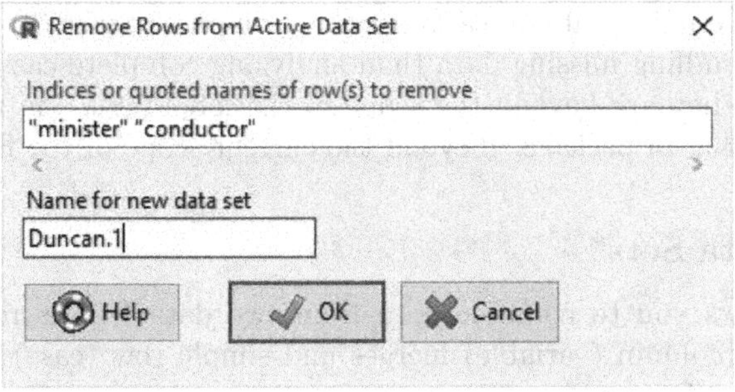

FIGURE 4.14: The *Remove Rows from Active Data Set* dialog.

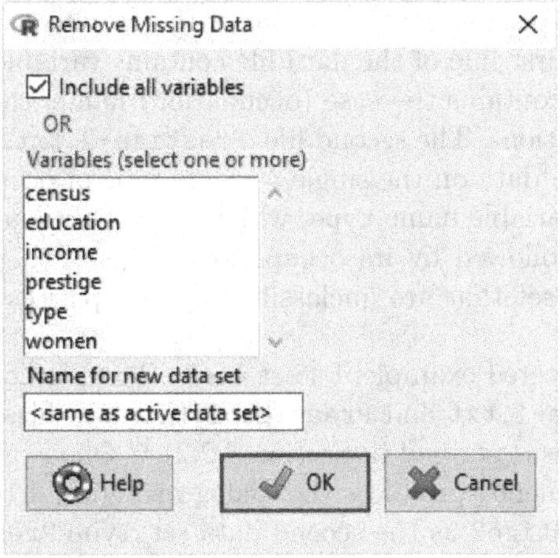

FIGURE 4.15: The *Remove Missing Data* dialog.

43 of the original 45 cases. Cases can be deleted by number as well as by name. For example, because `"minister"` and `"conductor"` are the 6th and 16th cases in the original `Duncan` data, I could have specified the cases to be deleted as `6 16`.

Selecting *Data > Active data set > Remove cases with missing data* produces the dialog shown in Figure 4.15, again for the `Prestige` data set from the **car** package. In completing the dialog, I leave the default *Include all variables* checked and retain the default name for the new data set, `<same as active data set>`. On clicking *OK*, the R Commander warns me that I'm replacing the existing `Prestige` data set, asking for confirmation. The new data set contains 98 of the 102 rows in the original `Prestige` data frame, eliminating the four occupations with missing `type`—none of the other variables in the original data set have missing values.

You may wish to remove missing data in this manner to analyze a consistent subset of complete cases. For example, suppose that you fit several regression models to the full `Prestige` data, and that some of the models include the variable `type` and others do not. The models that include `type` would be fit to 98 cases, and the others to 102 cases, making it inappropriate to compare the models (e.g., by likelihood ratio tests—see Section 7.7).

Two caveats: (1) Filtering missing data carelessly can needlessly eliminate cases. For example, if there's a variable in the data set that *you do not intend to use in your data analysis*, then it's not sensible to remove cases with missing data on this variable. You should only filter out cases with missing data on variables that you plan to use. (2) There are better general approaches to handling missing data than analyzing complete cases (see, e.g., Fox, 2016, Chapter 20), but they are beyond the scope of this book, and—in the absence of a suitable R Commander plug-in package—beyond the current scope of the R Commander.

4.5.3   Merging Data Sets*

The R Commander allows you to combine data from two data frames residing in the R workspace. Both simple column (variable) merges and simple row (case) merges are supported. I'll begin with the former.

To illustrate a column merge, I've divided the variables in the Canadian occupational prestige data into two plain-text, white-space-delimited files. The first file, Prestige-1.txt,[18] includes data on the numeric variables education, income, women, prestige, and census (see Table 4.2 on page 61 for definitions of the variables in the Canadian occupational prestige data). The first line of the data file contains variable names, and the first field in each subsequent line contains the case (occupation) name; there are thus 103 lines in this file, for the 102 occupations. The second file, Prestige-2.txt, is also a white-space-delimited, plain-text file, with data on the single variable type of occupation. The first line of the file contains only the variable name type, while the 98 subsequent lines each contain the name of an occupation followed by its occupational type (i.e., prof, wc, or bc); the four occupations in the data set that are unclassified by occupational type do not appear in Prestige-2.txt.

So as to provide an uncluttered example, I start a new R and R Commander session, and proceed to read the Prestige-1.txt and Prestige-2.txt data files into the data frames Prestige1 and Prestige2 (as described in Section 4.2.2).[19] Choosing *Data > Merge data sets* from the R Commander menus produces the dialog in Figure 4.16. I select Prestige1 as the first data set and Prestige2 as the second data set, type Prestige as the name of the merged data set replacing the default name MergedDataset, press the *Merge columns* radio button, and leave the *Merge only common rows or columns* box unchecked. Clicking *OK* merges the data sets, matching cases by row name, and producing a data set with 102 rows. The four cases that are present in Prestige1 but absent from Prestige2 have missing values (NA) for type. Had I checked the *Merge only common rows or columns* box, the merged data set would have included only the 98 cases present in *both* original data sets.

As demonstrated in this example, to merge variables from two data frames, the R Commander uses the row names of the data frames as the *merge key*. You may have to do some preliminary data management work on the two data sets to insure that their row names are consistent.

Row merges can also be performed via the *Merge Data Set* dialog by leaving the default *Merge rows* radio button pressed. If, as is typically the case, there are common variables in the two data sets to be merged, then these variables should have the same names in both data sets. You can choose to merge *only* variables that are common to both data sets or to merge *all* variables in each. In the latter event, variables that are in only one of the data sets will be filled out with missing values for the cases originating in the other data set.

To illustrate a row merge, I divided Duncan's occupational prestige data into three data files: Duncan-prof.txt, containing data for 18 professional, technical, and managerial occupations; Duncan-wc.txt, with data for 6 white-collar occupations; and Duncan-bc.txt

---

[18]Recall that all data files used in this book are available on the web site for book: See Section 1.5.

[19]Notice that I used the data set names Prestige1 and Prestige2 even though the corresponding data *files* have names containing hyphens (Prestige-1.txt and Prestige-2.txt): Hyphens aren't legal in R data set names.

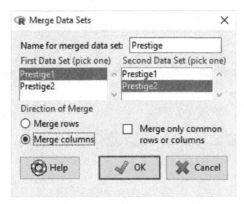

FIGURE 4.16: Using the *Merge Data Sets* dialog to combine variables from two data sets with some common cases.

with data for 21 blue-collar occupations. All three are plain-text, white-space-delimited files with variable names in the first line and case names in the first field of each subsequent line. All three files contain data for the variables type, income, education, and prestige (see Table 4.1 on page 60).

To merge the three parts of the Duncan data set, I begin by reading the partial data sets into the R Commander in the now-familiar manner, creating the data frames Duncan.bc, Duncan.wc, and Duncan.prof.[20] Next, I select *Data > Merge data sets* from the R Commander menus, and pick two of the three parts of the Duncan data, Duncan.bc and Duncan.wc, as illustrated at the left of Figure 4.17, creating the data frame Duncan. Finally, I perform another row merge, of Duncan and Duncan.prof, as shown at the right of Figure 4.17. Because I specify an existing data-set name (Duncan) for the merged data set, the R Commander asks me to confirm the operation. The result of the second merge is the complete Duncan data set with 45 cases and four variables, which is now the active data set in the R Commander.

---

[20] Again, the data sets must have legal R names that can't, for example, contain hyphens.

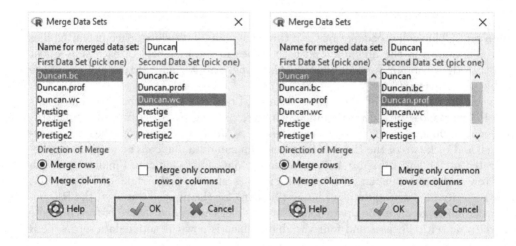

FIGURE 4.17: Using the *Merge Data Sets* dialog twice to merge three data sets by rows: Merging `Duncan.bc` and `Duncan.wc` to create `Duncan` (left); and then merging `Duncan` with `Duncan.prof` to update `Duncan` (right).

5

Summarizing and Graphing Data

This chapter explains how to use the R Commander to compute simple numerical summaries of data, to construct and analyze contingency tables, and to draw common statistical graphs. Most of the statistical content of the chapter is covered in a typical basic statistics course, although a few topics, such as quantile-comparison plots (in Section 5.3.1) and smoothing scatterplots (in Section 5.4.1), are somewhat more advanced.

Although most of the graphs produced by the R Commander use color, most of the figures in this chapter are rendered in monochrome.[1]

5.1 Simple Numerical Summaries

The R Commander *Statistics > Summaries* menu (see Figure A.4 on page 202) contains several items for summarizing data. I'll use the Canadian occupational prestige data (introduced in Section 4.2.3) to illustrate. This data set is most conveniently available in the `Prestige` data frame in the **car** package, which is one of the packages loaded when the R Commander starts up. I read the data via *Data > Data in packages > Read data set from an attached package* (as described in Section 4.2.4). Because the default alphabetic order of the levels of the `type` factor in the data set—`"bc"` (blue-collar), `"prof"` (professional and managerial), `"wc"` (white-collar)—is not the natural order, I reorder the levels of the factor with *Data > Manage variables in active data set > Reorder factor levels* (see Section 3.4).

Selecting *Statistics > Summaries > Active data set* produces the brief summary in Figure 5.1. There's a "five-number summary" for each numeric variable—reporting the minimum, first quartile, median, third quartile, and maximum of the variable—plus the mean, and the frequency distribution of the factor `type`, including a count of NAs.

Statistics > Summaries > Numerical summaries brings up the dialog box in Figure 5.2. I select the variables **education**, **income**, **prestige**, and **women** in the *Data* tab and retain the default choices in the *Statistics* tab. Clicking *OK* results in the output in Figure 5.3. Were I to press the *Summarize by groups* button in the *Data* tab, I could compute summary statistics separately for each level of **type**.

Choosing *Statistics > Summaries > Table of statistics* allows you to calculate a statistic for one or more numeric variables within levels or combinations of levels of one or more factors. To illustrate, I'll use the `Adler` data set from the **car** package. The data are from a social-psychological experiment, reported by Adler (1973), on "experimenter effects" in psychological research—that is, how researchers' expectations can influence the data that they collect. Adler recruited "research assistants," who showed photographs of individuals' faces to respondents; the respondents were asked by the research assistants to rate the apparent "successfulness" of the individuals in the photographs. In fact, Adler chose photographs that were average in their appearance of success, and the true subjects in the study were the

---

[1]Some figures in the chapter appear in color in the insert at the center of the book.

```
> summary(Prestige)
   education           income           women            prestige
 Min.   : 6.380   Min.   :   611   Min.   : 0.000   Min.   :14.80
 1st Qu.: 8.445   1st Qu.:  4106   1st Qu.: 3.592   1st Qu.:35.23
 Median :10.540   Median :  5930   Median :13.600   Median :43.60
 Mean   :10.738   Mean   :  6798   Mean   :28.979   Mean   :46.83
 3rd Qu.:12.648   3rd Qu.:  8187   3rd Qu.:52.203   3rd Qu.:59.27
 Max.   :15.970   Max.   : 25879   Max.   :97.510   Max.   :87.20
     census          type
 Min.   :1113    bc  :44
 1st Qu.:3120    wc  :23
 Median :5135    prof:31
 Mean   :5402    NA's: 4
 3rd Qu.:8312
 Max.   :9517
```

FIGURE 5.1: Summary output for the Prestige data set.

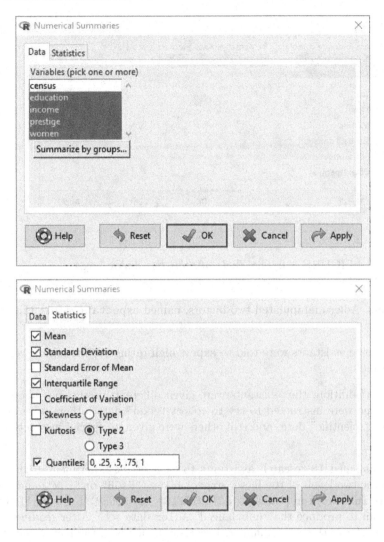

FIGURE 5.2: *Numerical Summaries* dialog box: *Data* tab (top) and *Statistics* tab (bottom).

```
> numSummary(Prestige[,c("education", "income", "prestige", "women")],
+    statistics=c("mean", "sd", "IQR", "quantiles"), quantiles=c(0,.25,.5,.75,1))
                  mean          sd        IQR     0%       25%      50%       75%
education     10.73804    2.728444     4.2025   6.38    8.4450    10.54   12.6475
income      6797.90196 4245.922227 4081.2500 611.00 4106.0000  5930.50 8187.2500
prestige      46.83333   17.204486    24.0500  14.80   35.2250    43.60   59.2750
women         28.97902   31.724931    48.6100   0.00    3.5925    13.60   52.2025
                  100%   n
education        15.97 102
income        25879.00 102
prestige         87.20 102
women            97.51 102
```

FIGURE 5.3: Numerical summaries for several variables in the Prestige data set.

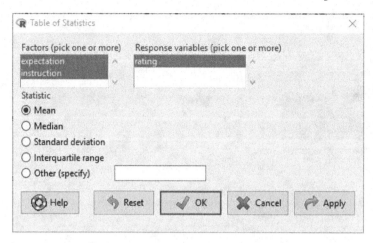

FIGURE 5.4: The *Table of Statistics* dialog box.

research assistants. Adler manipulated two factors, named `expectation` and `instruction` in the data set.

- `expectation`: Some assistants were told to expect high ratings, while others were told to expect low ratings.

- `instruction`: In addition, the assistants were given different instructions about how to collect data. Some were instructed to try to collect "good" data, others were instructed to try to collect "scientific" data, and still others were given no special instruction of this type.

Adler randomly assigned 18 research assistants to each of six experimental conditions—combinations of the two levels of the factor `expectation` (`"HIGH"` or `"LOW"`) and the three levels of the factor `instruction` (`"GOOD"`, `"SCIENTIFIC"`, or `"NONE"`). I deleted 11 of the 108 subjects at random to produce the "unbalanced" `Adler` data set.[2] After reading the data into the R Commander in the usual manner, I reorder the levels of the factor `instruction` from the default alphabetic ordering.

The *Table of Statistics* dialog box appears in Figure 5.4. I select both `expectation` and `instruction` in the *Factors* list box; because there's just one numeric variable in the data set—`rating`—it's preselected in the *Response variables* list box. The dialog includes radio buttons for calculating the mean, median, standard deviation, and interquartile range, along with an *Other* button, which allows you to enter any R function that computes a single number for a numeric variable. I retain the default *Mean*, and press the *Apply* button. Then, when the dialog reappears, I select *Standard deviation* and press *OK*. The output is shown in Figure 5.5. I'll defer interpreting the `Adler` data to Section 5.4 on graphing means and Section 6.1 on hypothesis tests for means.

Several of the *Statistics > Summaries* menu items and associated dialogs are very straightforward, and so, in the interest of brevity, I won't demonstrate their use here:[3]

---

[2]Of course, it isn't sensible to discard data like this, but I wanted to produce a more complex two-way analysis of variance example, with unequal numbers of cases in the combinations of levels of two factors; see Section 6.1.

[3]Two of the items in this menu, *Correlation test* and *Shapiro-Wilk test of normality*, perform simple hypothesis tests; I'll take them up in Chapter 6.

```
> with(Adler, tapply(rating, list(expectation, instruction), mean,
+   na.rm=TRUE))
          GOOD SCIENTIFIC  NONE
HIGH   4.066667  -6.944444 -10.0
LOW  -18.058824   1.923077  -3.5

> with(Adler, tapply(rating, list(expectation, instruction), sd, na.rm=TRUE))
        GOOD SCIENTIFIC     NONE
HIGH 16.64961   8.446874 15.63756
LOW  10.57397  11.715101 11.62781
```

FIGURE 5.5: Tables of means and standard deviations for `rating` in the `Adler` data set, classified by `expectation` and `instruction`.

- The *Frequency Distributions* dialog produces frequency and percentage distributions for factors, along with an optional chi-square goodness-of-fit test with user-supplied hypothesized probabilities for the levels of a factor.

- The *Count missing observations* menu item simply reports the number of NAs for each variable in the active data set.

- The *Correlation Matrix* dialog calculates Pearson product-moment correlations, Spearman rank-order correlations, or partial correlations for two or more numeric variables, along with optional pairwise p-values, computed with and without correction for simultaneous inference.

5.2 Contingency Tables

The *Statistics > Contingency tables* menu (see Figure A.4 on page 202) has items for constructing two-way and multi-way tables from the active data set. I demonstrated the *Two-Way Table* dialog in Section 3.5, and there is no need to repeat that demonstration here. Moreover, the *Multi-Way Table* dialog is similar, except that, in addition to selecting row and column factors for the contingency table, you can pick one or more "control" factors: A separate two-way partial table, optionally percentaged by rows or columns, is reported for each combination of levels of the control factors.

In contrast, the *Enter Two-Way Table* dialog (in Figure 5.6), selected via *Statistics > Contingency tables > Enter and analyze two-way table*, is unusual for the R Commander, in that it doesn't use the active data set. The dialog allows you to enter frequencies (counts) from an existing two-way contingency table, typically from a printed source such as a textbook. The sliders at the top of the *Table* tab control the number of rows and columns in the table. Initially, the table has 2 rows and 2 columns, and the cells of the table are empty.

Setting the sliders to 3 rows and 2 columns, I enter a frequency table taken from *The American Voter*, a classic study of electoral behavior by Campbell et al. (1960). The data originate in a panel study of the 1956 U.S. presidential election. During the campaign, survey respondents were asked how strongly (weak, medium, or strong) they preferred one candidate to the other, and after the election they were asked whether or not they had voted.

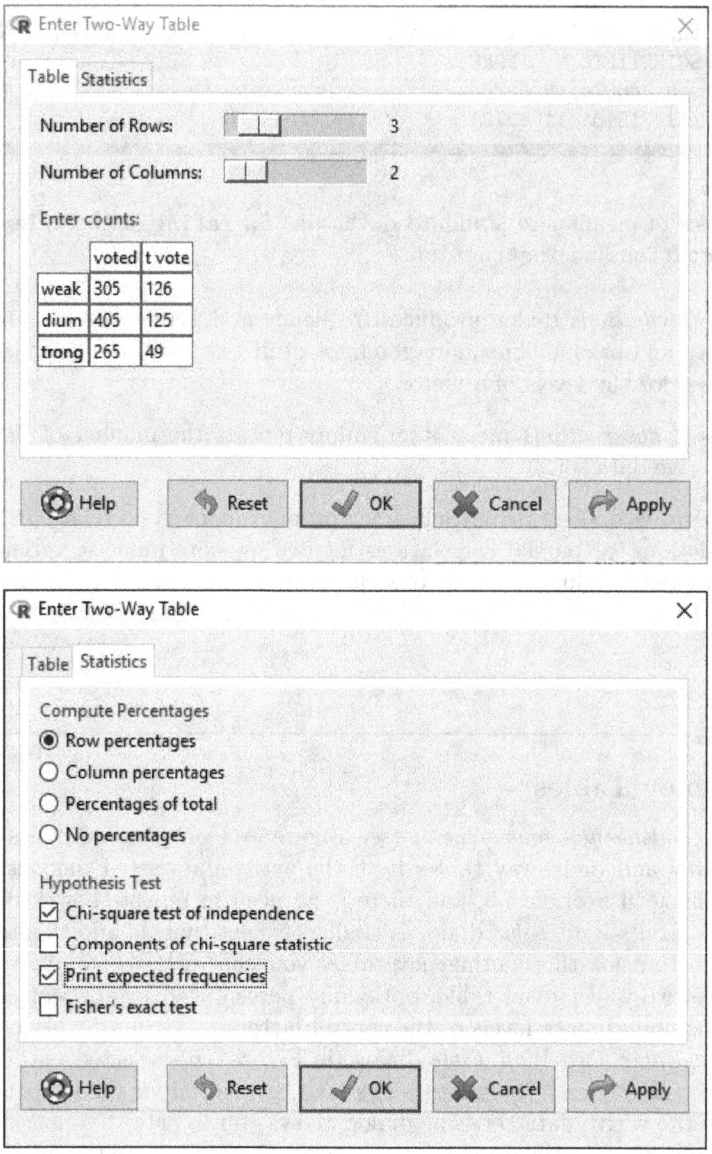

FIGURE 5.6: The *Enter Two-Way Table* dialog: *Table* tab (top) and *Statistics* tab (bottom).

```
> .Table <- matrix(c(305,126,405,125,265,49), 3, 2, byrow=TRUE)

> rownames(.Table) <- c('weak', 'medium', 'strong')

> colnames(.Table) <- c('voted', 'did not vote')

> .Table  # Counts
       voted did not vote
weak     305          126
medium   405          125
strong   265           49

> rowPercents(.Table) # Row Percentages
       voted did not vote Total Count
weak    70.8         29.2   100   431
medium  76.4         23.6   100   530
strong  84.4         15.6   100   314

> .Test <- chisq.test(.Table, correct=FALSE)

> .Test

        Pearson's Chi-squared test

data:   .Table
X-squared = 18.755, df = 2, p-value = 8.459e-05

> .Test$expected # Expected Counts
           voted did not vote
weak    329.5882    101.41176
medium  405.2941    124.70588
strong  240.1176     73.88235

> remove(.Test)

> remove(.Table)
```

FIGURE 5.7: Output produced by the *Enter Two-Way Table* dialog, having entered a contingency table from *The American Voter*.

The *Statistics* tab appears at the bottom of Figure 5.6. I check the box for *Row percentages* because the row variable in the table, intensity of preference, is the explanatory variable; the *Chi-square test of independence* checkbox is selected by default. I also check *Print expected frequencies*, which is *not* selected by default.

The output from the dialog is shown in Figure 5.7. Reported voter turnout increases with intensity of partisan preference, and the relationship between the two variables is highly statistically significant, with a very small p-value for the chi-square test of independence. All of the expected counts are much larger than necessary for the chi-square distribution to be a good approximation to the distribution of the test statistic; had that *not* been the case, a warning would have appeared, whether or not expected frequencies are printed.

5.3 Graphing Distributions of Variables

I'll use the Canadian occupational prestige data, read from the **car** package earlier in this chapter, to illustrate graphing distributions. There are, at this point in the chapter, two data sets in memory—the `Prestige` data set and the `Adler` data set—and the latter is the active data set. To change the active data set, I click on the *Data set* button in the R Commander toolbar and select `Prestige` in the resulting dialog.[4]

5.3.1 Graphing Numerical Data

The R Commander *Graphs* menu (see Figure A.6 on page 203) is divided into several groups of items, the second of which leads to dialogs for constructing graphs of the distribution of a numerical variable: *Index plot, Dot plot, Histogram*, nonparametric *Density estimate, Stem-and-leaf display, Boxplot*, and theoretical *Quantile-comparison plot*. Many of these graphs—specifically, dot plots, histograms, density estimates, and boxplots—can also show the distribution of a numeric variable within levels of (i.e., *conditional on*) a factor, and stem-and-leaf displays can be drawn back-to-back for the two levels of a dichotomous factor (see the example in Section 6.1.1).

Selecting *Graphs > Histogram* produces the dialog box in Figure 5.8. The *Data* tab, at the top of the figure, allows you to choose a numeric variable; I select `income`. Clicking the *Plot by groups* button brings up the *Groups* sub-dialog shown at the center of the figure; because there is only one factor in the data set, `type`, it is preselected. Clicking *OK* in the *Groups* sub-dialog returns to the main dialog, and now the *Plot by* button reads *Plot by: type*. The *Options* tab is at the bottom of Figure 5.8. Leaving all of the options at their defaults and clicking *OK* produces the vertically aligned histograms in Figure 5.9.

If you don't like the default number of bins, which results from leaving the *Number of bins* text box at <auto>, you can type a target number for the number of bins:[5] As a general matter, as you increase the number of bins, the width of each bin decreases. You can conveniently experiment with the number of bins by pressing the *Apply* button rather than the *OK* button in the dialog.

The dialogs for the other distributional displays differ only in their *Options* tabs and whether or not (as noted above) they support plotting by groups. Figure 5.10 shows the

---

[4]Although the R Commander can switch among data sets in this manner, you'll no doubt work with a single data set in most of your R Commander sessions.

[5]The number of bins specified is just a *target* because the program that creates the histogram also tries to use "nice" numbers for the boundaries of the bins. The default target number of bins is determined by Sturges's rule (Sturges, 1926).

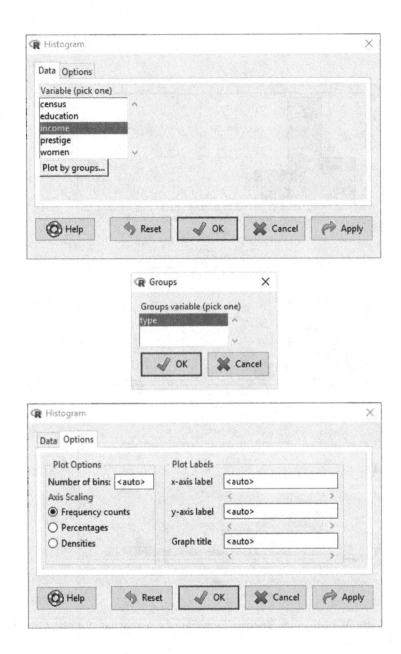

FIGURE 5.8: *Histogram* dialog, showing the *Data* tab (top), *Groups* sub-dialog (center), and *Options* tab (bottom).

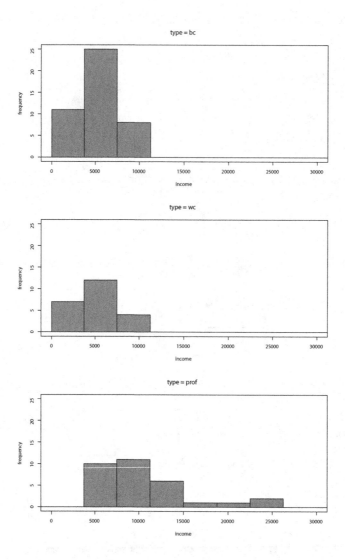

FIGURE 5.9: Histograms of average `income` by `type` of occupation, for the Canadian occupational prestige data.

default distributional displays for education in the Canadian occupational prestige data set.[6] There is also a "rug plot" at the bottom of the density estimate (center-right panel), showing the location of the data values. By default the quantile-comparison plot (lower-right) compares the distribution of the data to the normal distribution, but you can also plot against other theoretical distributions.[7]

In the index plot (at the upper-left) and quantile-comparison plot (at the lower-right), the two most extreme values are automatically identified by default, but because these values are close to each other in the graphs, the labels for the points are over-plotted. The case labels are also displayed, however, in the R Commander *Output* pane (not shown), and they are university.teachers and physicians.

The default stem-and-leaf display for education appears in Figure 5.11; it is text output and so is printed in the *Output* pane.

---

[6]To obtain the marginal histogram for education—that is, *not* to plot by occupational type—either press the *Reset* button in the *Histogram* dialog, or press the *Plot by: type* button, and deselect type in the resulting *Groups* sub-dialog by *Ctrl*-clicking on it in the *Groups variable* list.

[7]Probability distributions are discussed in Chapter 8.

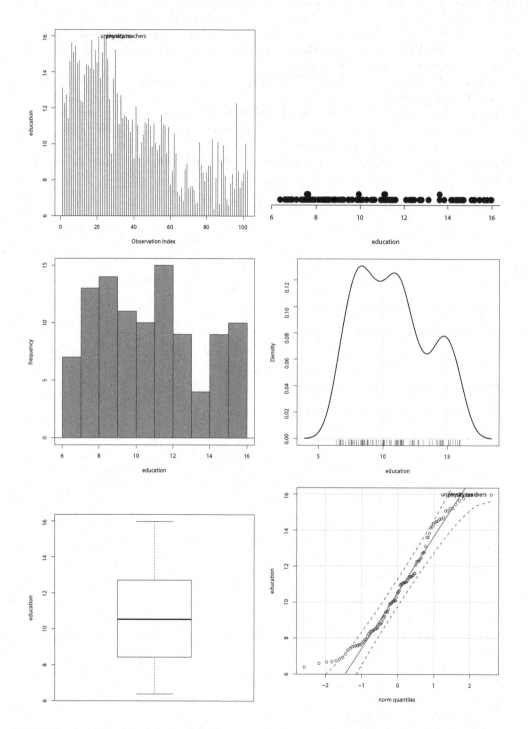

FIGURE 5.10: Various default distributional displays for average `education` in the Canadian occupational prestige data. From top to bottom and left to right: index plot, dot plot, histogram, nonparametric density estimate with rug plot, boxplot, and quantile-comparison plot comparing the distribution of `education` to the normal distribution.

```
> with(Prestige, stem.leaf(education, na.rm=TRUE))
1 | 2: represents 1.2
 leaf unit: 0.1
            n: 102
    1     6* | 3
    7     6. | 666789
   10     7* | 134
   20     7. | 5555667899
   27     8* | 1233444
   34     8. | 5567788
   40     9* | 012444
   44     9. | 6899
   50    10* | 000012
  (4)    10. | 5569
   48    11* | 00000112334444
   34    11. | 56
   32    12* | 022334
   26    12. | 777
   23    13* | 1
   22    13. | 668
   19    14* | 1234
   15    14. | 55667
   10    15* | 00224
    5    15. | 67999
```

FIGURE 5.11: Default "Tukey-style" stem-and-leaf display for `education` in the Canadian occupational prestige data. The column of numbers to the left of the stems represents "depths"—counts in to the median from both ends of the distribution—with the parenthesized value (4) giving the count for the stem containing the median. Note the divided stems, with *x*. stems containing leaves 0–4 and *x*∗ stems leaves 5–9. Five-part stems are similarly labelled *x*. with leaves 01, *x*t with leaves 23, *x*f with leaves 45, *x*s with leaves 67, and *x*∗ with leaves 89.

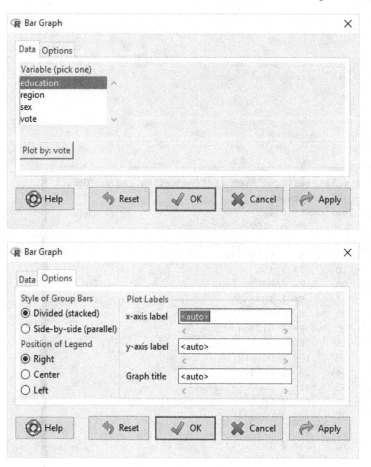

FIGURE 5.12: *Bar Graph* dialog, showing the *Data* tab (top) and *Options* tab (bottom). I previously pressed the *Plot by* button and selected the factor vote.

5.3.2   Graphing Categorical Data

I'll demonstrate graphing the distribution of a categorical variable by using the Chile data set from the **car** package. This data set is from a poll conducted about six months before the 1988 Chilean plebiscite on the continuation of military rule: voting "yes" in the plebiscite represented support for Pinochet's military government, while "no" represented support for a return to electoral democracy. Two of the variables in the Chile data set are the factors vote, with levels "N" (no), "Y" (yes), "U" (undecided), and "A" (abstain), and education, with levels "P" (primary), "S" (secondary), and "PS" (post-secondary). In both cases, the default alphabetic ordering of the factor levels isn't the natural ordering, and so, after reading the data, I change the orderings via *Data > Manage variables in active data set > Reorder factor levels* (see Section 3.4).

The *Graphs* menu includes two simple distributional plots for factors: frequency bar graphs and pie charts. Because it allows for dividing bars by the value of a second factor, the *Bar Graph* dialog, shown in Figure 5.12, is the more complex of the two. In the *Data* tab, at the top of the figure, I select the factor education to define the bars. I previously pressed the *Plot by* button and chose vote in the resulting *Groups* sub-dialog, and so the button displays *Plot by: vote*. I retain all of the default choices in the *Options* tab at the

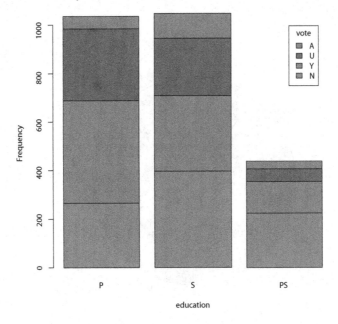

FIGURE 5.13: Bar graph for education in the Chilean plebiscite data, with bars divided
by vote. A color version of this figure appears in the insert at the center of the book.

bottom of Figure 5.12. Clicking *OK* produces the graph in Figure 5.13. It's apparent that
relative support for the military government declined with education, but that overall the
plebiscite appeared close (visually summing and comparing the "N" and "Y" areas across
the bars).

Overall voting intentions are displayed in the pie chart in Figure 5.14. The *Pie Chart*
dialog, not shown, simply allows you to pick a factor and, optionally, provide axis labels
and a graph title.

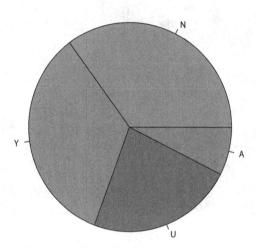

FIGURE 5.14: Pie chart for vote in the Chilean plebiscite data. A color version of this figure appears in the insert at the center of the book.

5.4   Graphing Relationships

The third section of the *Graphs* menu is for graphing relationships between and among variables, including *scatterplots*, *scatterplot matrices*, and *3D scatterplots* for numeric variables, *line plots*, which are typically for time series data, *plots of means* of a numeric variable classified by one or more factors, *strip charts*, which are similar to conditional dot plots (discussed in Section 5.3.1), and *conditioning plots*, which are capable of representing the relationships between one or more numeric response variables and explanatory variables that are both numeric and factors.[8] I'll focus here on scatterplots for two numeric variables, scatterplot matrices for several numeric variables, 3D scatterplots for three numeric variables, and plots of means of a numeric variable classified by one or two factors.

In addition, and as mentioned previously, some of the distributional graphs discussed in Section 5.3.1 can be used to examine the relationship between a numeric response variable and a factor. These include dot plots, histograms, stem-and-leaf displays (with a dichotomous factor), and boxplots.

5.4.1   Simple Scatterplots

To illustrate the construction of scatterplots, scatterplot matrices, and 3D scatterplots, I return to the Canadian occupational prestige data in the previously read Prestige data set. Choosing *Graphs > Scatterplot* from the R Commander menus brings up the dialog box in Figure 5.15. As you can see, there are many options in the dialog, some of which I'll

---

[8]The R Commander dialogs for strip charts and conditioning plots were originally contributed by Richard Heiberger.

describe presently. In the *Data* tab, I select `income` (which is the explanatory variable) as the *x-variable* and `prestige` (the response variable) as the *y-variable*. I retain all of the defaults in the *Options* tab, clicking *Apply* to draw the simple scatterplot in Figure 5.16. Occupational `prestige` apparently increases with `income`, but the relationship is nonlinear, with the rate of increase declining with `income`.

5.4.2 Enhanced Scatterplots*

To draw the scatterplot in Figure 5.17, I click on the *Plot by groups* button in the *Data* tab; because it's the only factor in the data set, `type` is preselected in the resulting *Groups variable* sub-dialog (not shown). The sub-dialog also has a checkbox for plotting lines by group, which is selected by default. In the *Options* tab, I check the boxes for *Least-squares line*, *Smooth line*, and *Plot concentration ellipses*. I also change the *Legend Position* from the default *Above plot* to *Bottom right*.

The smooth line is produced by a method of *nonparametric regression* called *loess*, an acronym for *local regression*, which traces how the average value of y changes with x without assuming that the relationship between y and x takes a specific form. The *span* of the loess smoother is the percentage of the data used to compute each smoothed value: The larger the span, the smoother the resulting loess regression. The default span is 50%, a value that I increase to 100% because of the small number of cases in each level of occupational `type`. As a general matter, you want to select the smallest span that produces a reasonably smooth regression, a value that you can determine by trial and error, pressing the *Apply* button in the dialog each time you adjust the *Span* slider.

Concentration ellipses are summaries of the variational and correlational structure of the points. For bivariately normally distributed data, concentration ellipses enclose specific fractions of the data—50% and 90% by default; the ellipses are computed robustly, however, to reduce the impact of outliers. To avoid an overly cluttered graph, I set the *Concentration levels* to 0.5, to draw only one ellipse for each occupational `type`.

The scatterplot in Figure 5.17 suggests that the apparently nonlinear relationship between `prestige` and `income` is due to occupational `type`: Within levels of `type`, the relationship is reasonably linear, but with the slope changing across levels.

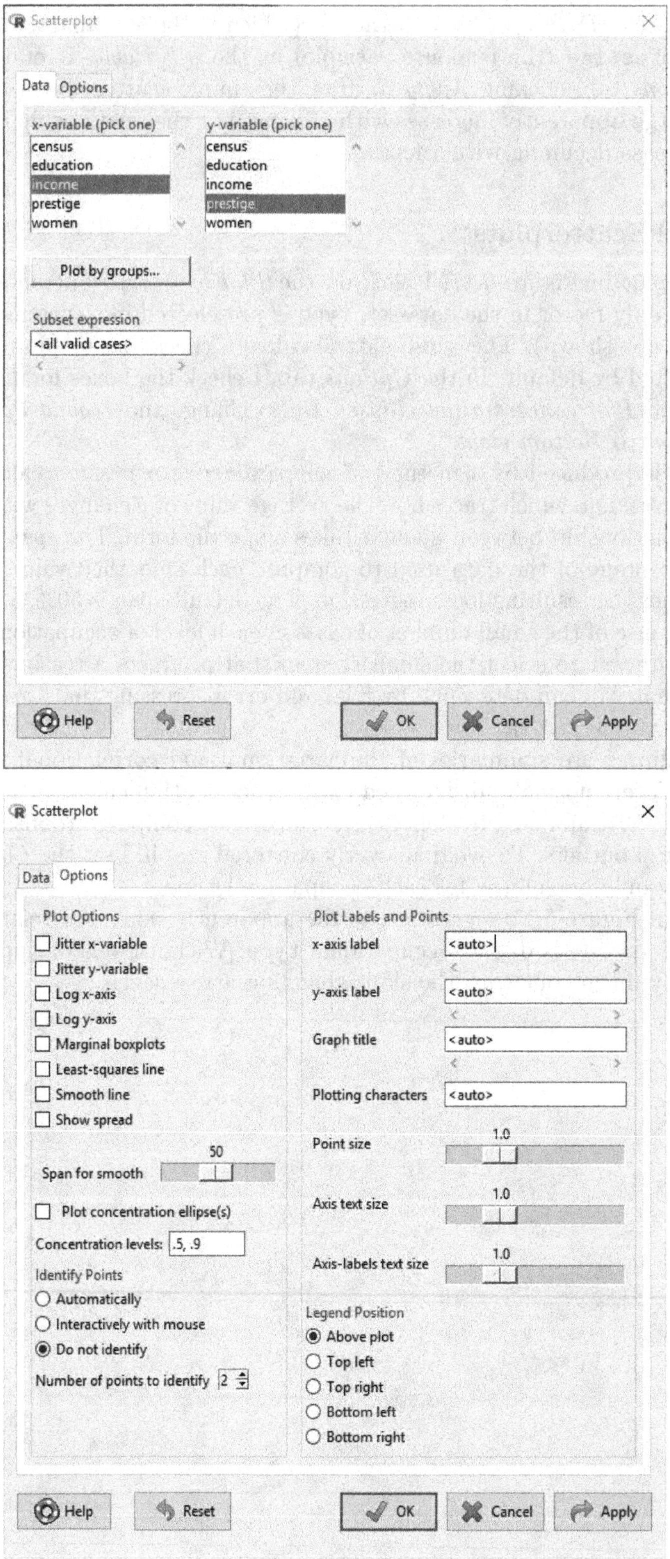

FIGURE 5.15: *Scatterplot* dialog: *Data* tab (top) and *Options* tab (bottom).

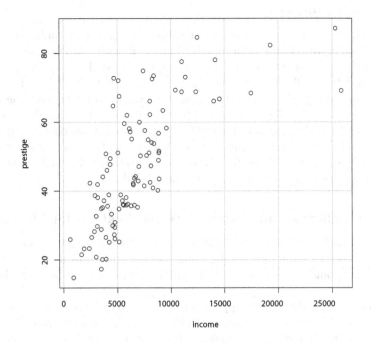

FIGURE 5.16: Simple scatterplot of **prestige** vs. **income** for the **Prestige** data.

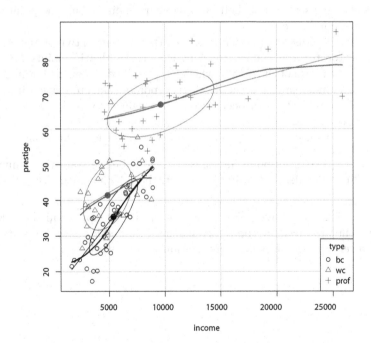

FIGURE 5.17: Enhanced scatterplot of **prestige** vs. **income** by occupational **type**, showing 50% concentration ellipses, least-squares lines, and loess lines. A color version of this figure appears in the insert at the center of the book.

5.4.3   Scatterplot Matrices

A *scatterplot matrix* displays the pairwise relationships among several numeric variables; it is the graphical analog of a correlation matrix. The *Scatterplot Matrix* dialog, shown in Figure 5.18, is similar in most respects to the *Scatterplot* dialog. I select several variables in the *Data* tab and leave all of the choices in the *Options* tab at their defaults. Each off-diagonal panel in the resulting scatterplot matrix in Figure 5.19 displays the pairwise scatterplot for two variables, while the diagonal panels show the marginal distributions of the variables. The plots in the first row, for example, have education on the vertical axis, while those in the first column have education on the horizontal axis—and similarly for the other variables in the graph. Thus, the scatterplot in the second row, first column has income on the vertical axis and education on the horizontal axis.

5.4.4   Point Identification in Scatterplots and Scatterplot Matrices

Both the *Scatterplot* and the *Scatterplot Matrix* dialogs provide an option for automatic identification of noteworthy cases. Automatic point identification uses a robust method to find the most unusual points in each scatterplot, with the number of points to be identified set by the user.

The *Scatterplot* dialog additionally supports *interactive* point identification, selected by pressing the corresponding radio button in the *Options* tab. Under Windows or Linux/Unix, interactive point identification displays a message box with the text *Use left mouse button to identify points, right button to exit*; under Mac OS X, the message reads *Use left mouse button to identify points, esc key to exit*. In either case, click the *OK* button to dismiss the message box. On all operating systems, the mouse cursor turns into "cross-hairs" (a +) when the cursor is over the scatterplot. Left-clicking near a point labels the point with the row name of the corresponding case.

To produce Figure 5.20, I use the *Scatterplot* dialog (not repeated) to plot income (on the *y*-axis) versus education (on the *x*-axis), checking the box to identify points *Interactively with the mouse*. I click near two of the points, which are identified as general.managers and physicians. These are, incidentally, precisely the two points that are flagged if automatic point identification is employed. Both occupations have unusually high values of income given their levels of education.

There are two issues to keep in mind about interactive point identification:

1. It is necessary to exit from point identification mode before you can do anything else in the R Commander. If you forget to exit, the R Commander will appear to freeze!

2. Scatterplots using interactive point identification do not appear in the R Markdown document produced by the R Commander (see Section 3.6). As in the example, however, automatic point identification usually works quite well.

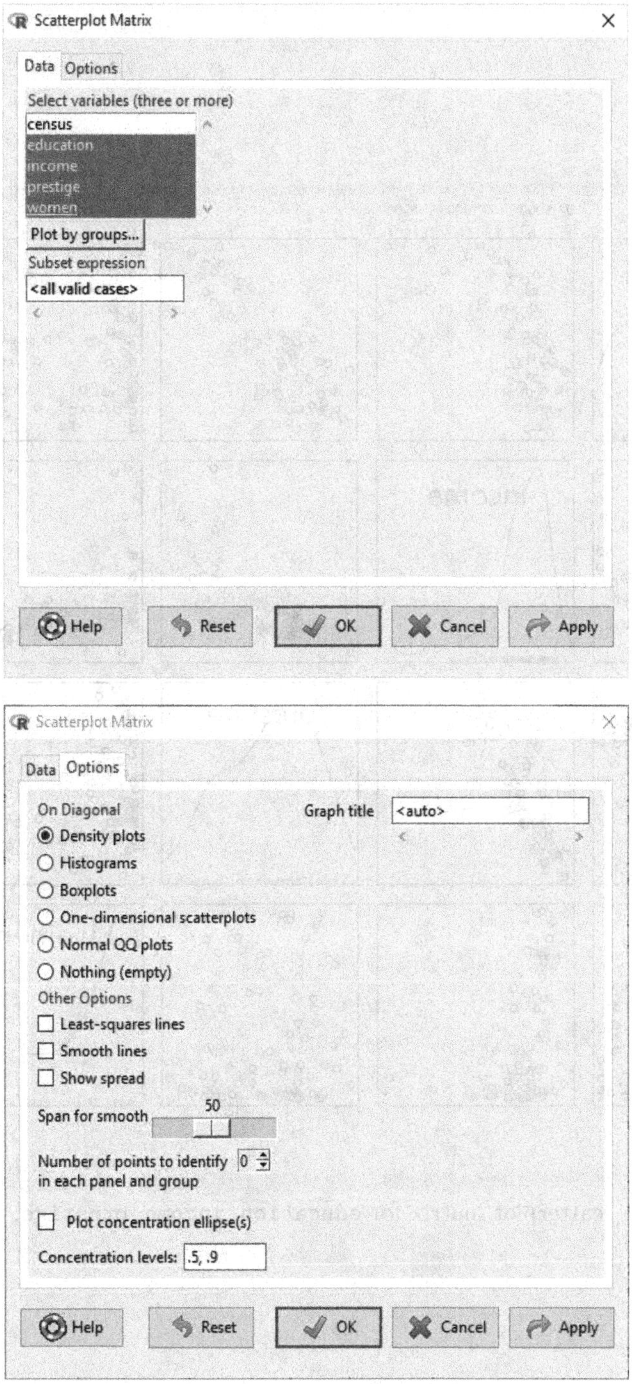

FIGURE 5.18: *Scatterplot Matrix* dialog, with *Data* tab (top) and *Options* tab (bottom).

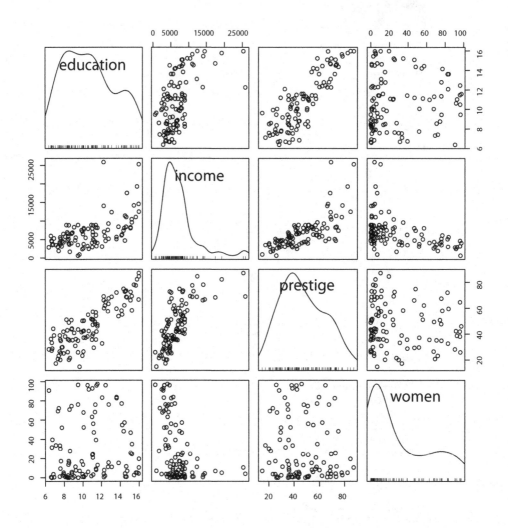

FIGURE 5.19: Default scatterplot matrix for education, income, prestige, and women in the Prestige data set.

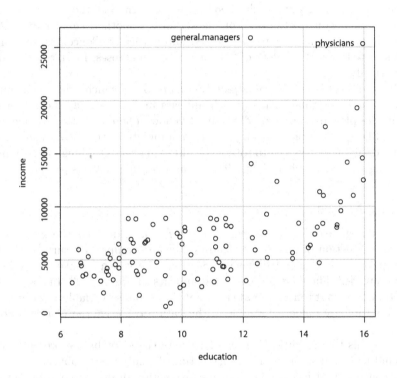

FIGURE 5.20: Scatterplot for `education` and `income` in the `Prestige` data set, with two points identified interactively by mouse clicks.

5.4.5   3D Scatterplots*

You can draw a dynamic three-dimensional scatterplot for three numeric variables by choosing *Graphs > 3D graph > 3D scatterplot* from the R Commander menus, bringing up the dialog box in Figure 5.21. The structure of the *3D Scatterplot* dialog is very similar to that of the *Scatterplot* dialog.

In the *Data* tab, I select two numeric explanatory variables, education and income, along with a numeric response variable, prestige. As in a 2D scatterplot, I could (but don't) use the factor type to plot by groups—for example, using different colors for points in the various levels of type.

In addition to the default selections in the *Options* tab, I opt to plot the least-squares regression plane and an additive nonparametric regression, which allows nonlinear partial relationships between prestige and each of education and income; the degrees of freedom for these terms are analogous to the span of the 2D loess smoother: smaller *df* produce a smoother fit to the data. I allow *df* to be selected automatically. Concentration ellipsoids (which are not selected) are 3D analogs of 2D concentration ellipses. I also elect to identify 2 points automatically.[9]

Clicking *OK* produces the 3D scatterplot in Figure 5.22, which appears in an *RGL device* window.[10] In the original image (which appears in the insert at the center of the book), points in the plot are represented by small yellow spheres. A static image doesn't do justice to the 3D dynamic plot, which you can manipulate in the *RGL device* window: Left-clicking and dragging allows you to rotate the plot—in effect grabbing onto an invisible sphere surrounding the data—while right-clicking and dragging changes perspective.

5.4.6   Plotting Means

The *Plot Means* dialog, selected by *Graphs > Plot of means*, displays the mean of a numeric variable as a function of one or two factors. For an example, I return to the Adler experimenter-expectations data (introduced in Section 5.1), which I make the active data set in the R Commander. The *Plot Means* dialog appears in Figure 5.23. In the *Data* tab, I select the factors expectation and instruction; the response variable rating is preselected because it's the only numeric variable in the data set. I leave the *Options* tab in its default state.

Clicking *OK* yields the graph in Figure 5.24, where the error bars represent ±1 standard error around the means. Apparently, instructing the subjects to obtain "good" data produces a bias consistent with expectation, while instructing subjects to obtain "scientific" data or providing no instruction produces a smaller bias in the opposite direction.

---

[9]Interactive point identification in 3D works differently than in a 2D scatterplot: You right-click and drag a box around the point or points you want to identify, repeating this procedure as many times as you want. To exit from point identification mode, you must right-click in an area of the plot where there are no points. You can identify points interactively by checking the appropriate box in the *3D Scatterplot Options* tab or, after drawing a 3D scatterplot, by selecting *Graphs > 3D graph > Identify observations with mouse* from the R Commander menus.

[10]The R Commander uses the scatter3d function in the **car** package to draw 3D scatterplots; scatter3d, in turn, employs facilities provided by the **rgl** package (Adler and Murdoch, 2015) for constructing 3D dynamic graphs.

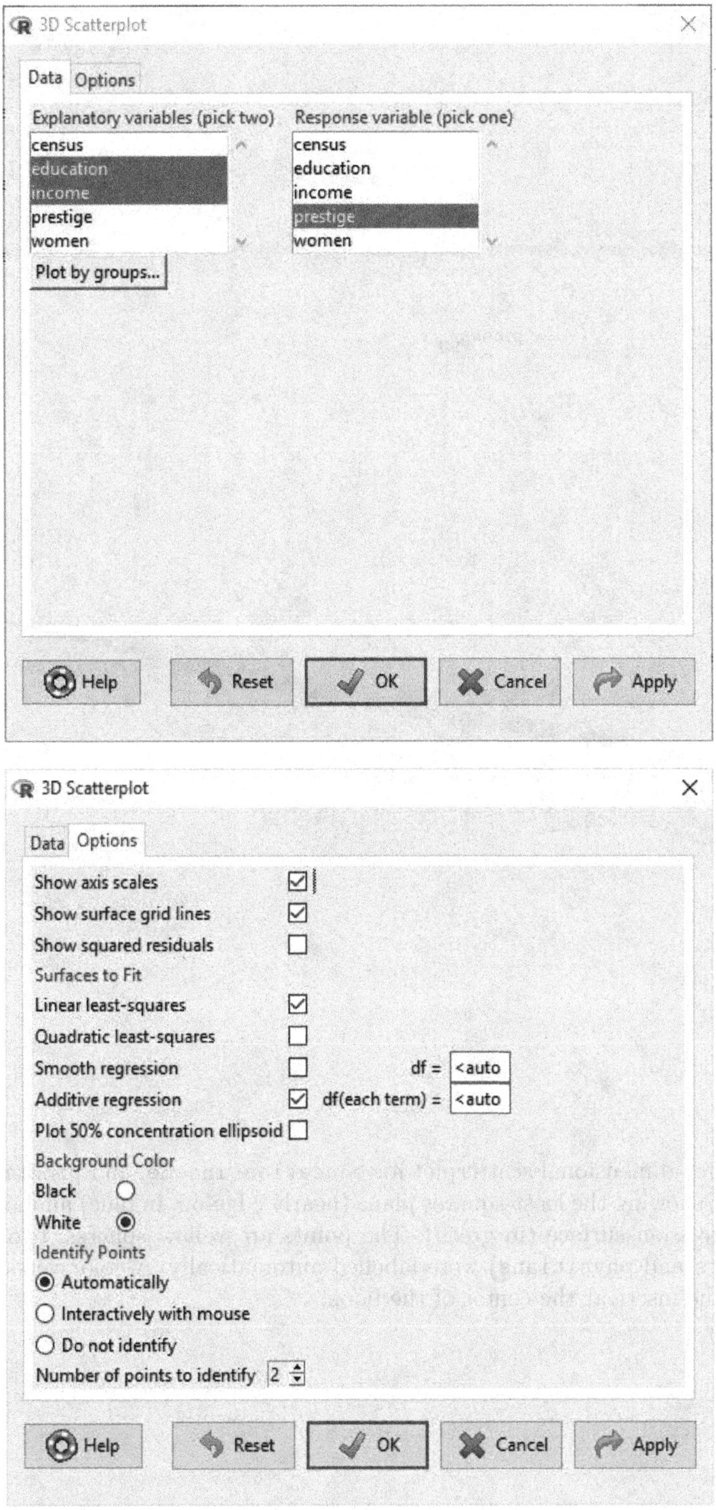

FIGURE 5.21: The *3D Scatterplot* dialog: *Data* tab (top) and *Options* tab (bottom).

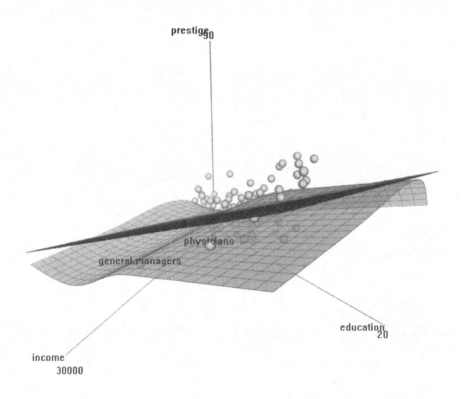

FIGURE 5.22: Three-dimensional scatterplot for education, income, and prestige in the Prestige data set, showing the least-squares plane (nearly edge-on, in blue) and an additive nonparametric regression surface (in green). The points are yellow spheres, two of which (general.managers and physicians) were labelled automatically. A color version of this figure appears in the insert at the center of the book.

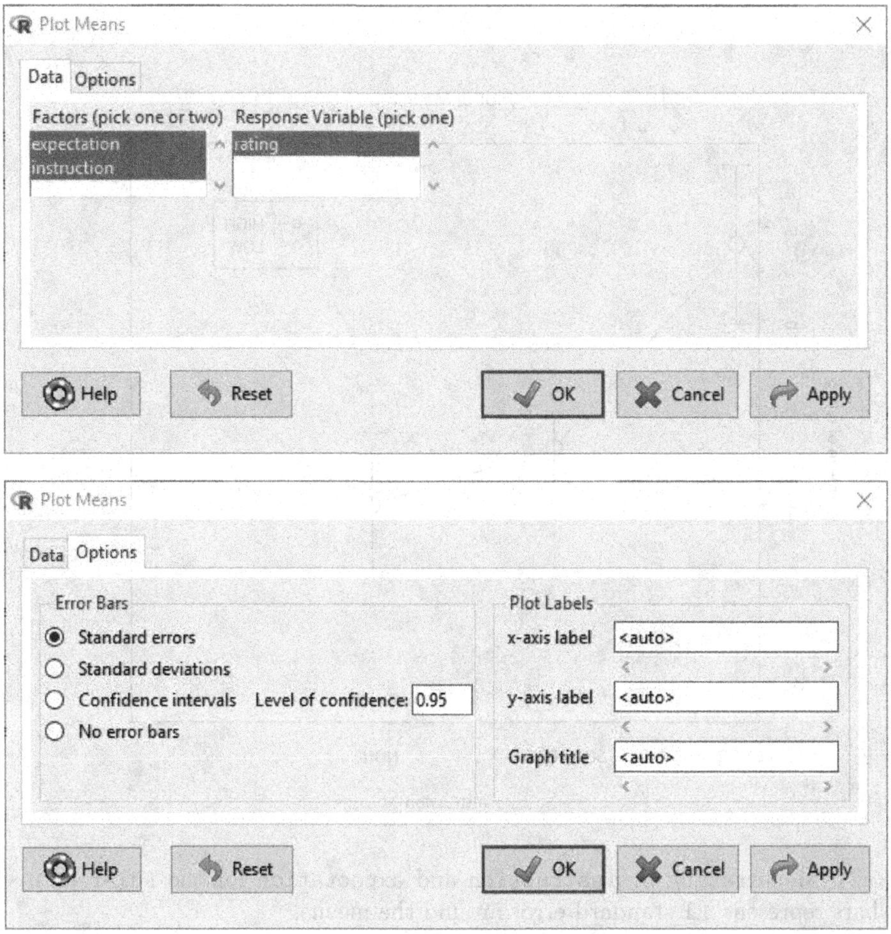

FIGURE 5.23: *Plot Means* dialog box: *Data* tab (top) and *Options* tab (bottom).

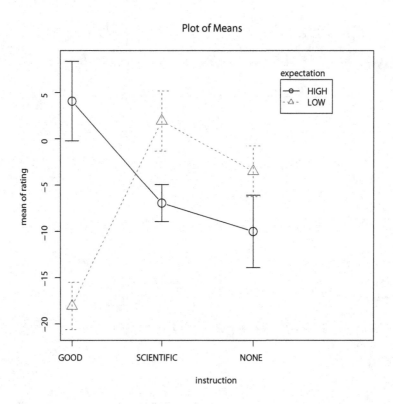

FIGURE 5.24: Mean **rating** by **instruction** and **expectation** for the **Adler** data set. The error bars represent ±1 standard error around the means.

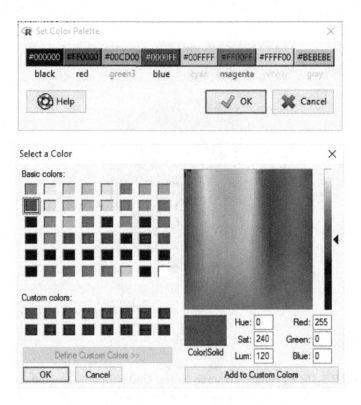

FIGURE 3.25: *Set Color Palette* dialog (top) and the Windows *Select a Color* sub-dialog (bottom).

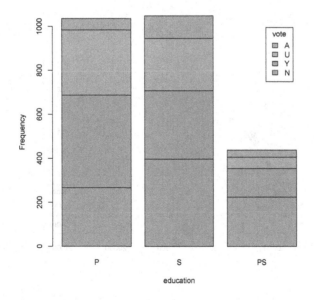

FIGURE 5.14: Bar graph for `education` in the Chilean plebiscite data, with bars divided by `vote`.

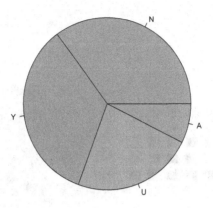

FIGURE 5.15: Pie chart for `vote` in the Chilean plebiscite data.

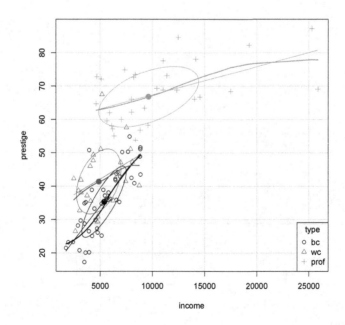

FIGURE 5.17: Enhanced scatterplot of `prestige` vs. `income` by occupational `type`, showing 50% concentration ellipses, least-squares lines, and loess lines.

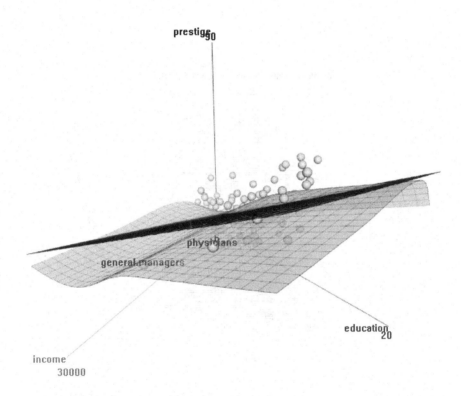

FIGURE 5.22: Three-dimensional scatterplot for education, income, and prestige in the Prestige data set, showing the least-squares plane (nearly edge-on, in blue) and an additive nonparametric-regression surface (in green). The points are yellow spheres, two of which (general.managers and physicians) were labelled automatically.

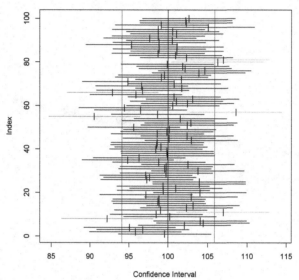

FIGURE 9.4: Ninety-five percent confidence intervals based on the means $\bar{x}$ of 100 samples, each of size $n = 25$, drawn from a normal population with mean $\mu = 100$ and (known) standard deviation $\sigma = 15$.

6

Simple Statistical Tests

This chapter shows how to compute simple statistical hypothesis tests and confidence intervals for means, for proportions, and for variances, along with simple nonparametric tests, a test of normality, and correlation tests. Many of these tests are typically taken up in a basic statistics class, and, in particular, tests and confidence intervals for means and proportions are often employed to introduce statistical inference.

6.1 Tests on Means

The R Commander *Statistics > Means* menu (see Figure A.4 on page 202) includes items for tests for a single mean, for the difference between means from independent samples, for the difference between means from matched (paired) samples, and for one-way and multi-way analysis of variance (ANOVA). The test for matched pairs compares the means of two variables in the active data set, presumably measured on the same scale (e.g., husband's and wife's annual income in dollars in a data set of married heterosexual couples).

6.1.1 Independent Samples Difference-of-Means *t* Test

The dialogs for tests on a single mean and for the difference between two means are similar, and so I'll illustrate with an *independent-samples t test*, using the **Guyer** data set in the **car** package. I begin by reading the data via *Data > Data in packages > Read data set from an attached package* (as described in Section 4.2.4).

The **Guyer** data set is drawn from an experiment, reported by Fox and Guyer (1978), in which 20 four-person groups each played 30 trials of a "prisoners' dilemma" game. On each trial of the experiment, the individuals in a group could make cooperative or competitive choices, and the response variable in the data set (**cooperation**) is the number of cooperative choices, out of the 120 choices made in each group during the course of the experiment.

Ten of the groups were randomly assigned to a treatment in which their choices were "public," in the sense that each individual's choice on a trial was visible to the other members of the group, while the other 10 groups made their choices anonymously, so that group members were aware only of the *numbers* of cooperative and competitive choices made on each trial. The treatment for each group is recorded in the factor **condition** as P (public choice) or **A** (anonymous choice).[1]

With **Guyer** as the active data set, selecting *Statistics > Means > Independent samples t-test* brings up the dialog box in Figure 6.1. The *Groups* list box in the *Data* tab displays the two-level factors in the data set; I select **condition**. The *Response Variable* list box

---

[1]There is a third variable in the **Guyer** data set, **sex**, indicating the gender composition of each group, with half the groups composed of females (coded F) and the other half of males (coded M). This factor won't figure in the *t* test that I report in this section, but I invite the reader to perform a two-way analysis of variance for the **Guyer** data, as described in Section 6.1.3.

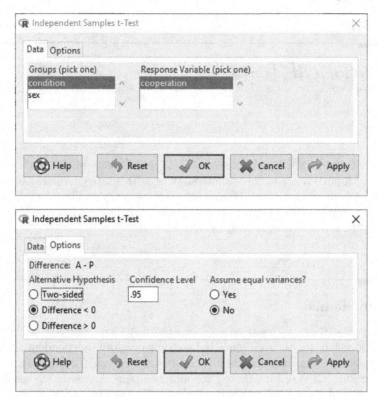

FIGURE 6.1: *Independent Samples t-Test* dialog box, showing the *Data* and *Options* tabs.

includes numeric variables in the active data set; because there is just one numeric variable—`cooperation`—it is preselected. In this example, there are equal numbers of cases—10 each—in the two groups to be compared, but equal sample sizes aren't necessary for a difference-of-means t test.

The levels of `condition` are in default alphabetic order, and so the *Options* tab shows that the difference in means will be computed as A − P (i.e., anonymous minus public choice, $\bar{x}_A - \bar{x}_P$). Because I expect higher cooperation in the public-choice condition, I select a directional alternative hypothesis—that the difference in "population" means between the anonymous and public-choice conditions is less than 0 (i.e., negative): H_a: $\mu_A - \mu_P < 0$; the default choice is a *Two-sided* alternative hypothesis: H_a: $\mu_A - \mu_P = 0$. I leave the other options at their defaults: a 95% confidence interval, and no assumption of equal variances for the two groups.

Because the alternative hypothesis is directional, the `t.test` command invoked by the dialog also computes a *one-sided confidence interval* for the difference in means.[2] Because equal group variances are not assumed, `t.test` computes the standard error of the difference in means using separate group standard deviations and approximates the degrees of freedom for the test by the *Welch–Satterthwaite* formula (see, e.g., Moore et al., 2013, p. 480).

---

[2]The lower bound of the one-sided confidence interval is necessarily $-\infty$. If, as is likely, you're unfamiliar with one-sided confidence intervals, you can simply disregard the reported confidence interval when you perform a one-sided test.

```
> t.test(cooperation~condition, alternative='less', conf.level=.95,
+   var.equal=FALSE, data=Guyer)

        Welch Two Sample t-test

data:  cooperation by condition
t = -2.6615, df = 15.237, p-value = 0.0088
alternative hypothesis: true difference in means is less than 0
95 percent confidence interval:
      -Inf -5.061611
sample estimates:
mean in group A mean in group P
        40.9            55.7
```

FIGURE 6.2: One-sided independent-samples difference-of-means t test for `cooperation` by `condition` in the `Guyer` data set.

The resulting output, in Figure 6.2, shows that the mean number of cooperative choices is indeed higher in the public-choice group ($\bar{x}_P = 55.7$) than in the anonymous-choice group ($\bar{x}_A = 30.9$), and that this difference is statistically significant ($p = 0.0088$, one-sided). The `t.test` function doesn't report the standard deviations in the two groups, but they are simple to compute via *Statistics > Summaries > Numerical summaries* (as described in Section 5.1) as $s_P = 14.85$ and $s_A = 9.42$.

The difference-of-means t test makes the assumption that the response variable is normally distributed within groups in the "population," and, because the sample sizes are small in the `Guyer` data set, I should be concerned if the distributions are skewed or if there are outliers. To check, I draw back-to-back stem-and-leaf displays for `cooperation` by `condition`, selecting *Graphs > Stem-and-leaf display* from the R Commander menus (see Section 5.3.1), taking all of the defaults in the resulting dialog (except for plotting back-to-back), and producing the result shown in Figure 6.3. There are no obvious problems here, although it's apparent that the values for the public-choice group are more variable than those for the anonymous-choice group. It would indeed have been more sensible to examine the data *before* performing the t test!

6.1.2   One-Way Analysis of Variance

One-way analysis of variance tests for differences among the means of several independent samples. To illustrate, I'll use data from a memory experiment conducted by Friendly and Franklin (1980). Subjects in the experiment were presented with a list of 40 words that they were asked to memorize. After a distracting task, the subjects recalled as many words as they could on each of five trials. Thirty subjects were assigned randomly to one of three experimental conditions, 10 to each condition: In the *standard free recall* condition, the words were presented in random order on each trial; in the *before* condition, words that were recalled on the previous trial were presented first to the subject in the order in which they were recalled, followed by the forgotten words; in the *meshed* condition, recalled words were also presented in the order in which they were previously recalled, but were intermixed with the forgotten words. As in the preceding t test example, there are equal numbers of cases in the groups (here, three groups), but the one-way ANOVA procedure in the R Commander can also handle unequal sample sizes.

```
> with(Guyer, stem.leaf.backback(cooperation[condition == "A"],
+   cooperation[condition == "P"], na.rm=TRUE))
_____
  1 | 2: represents 12, leaf unit: 1
cooperation[condition == "A"]
                  cooperation[condition == "P"]
_____
          | 2* |
  1      7| 2. |9        1
  3     40| 3* |
  4      9| 3. |7        2
 (4)  4410| 4* |
          | 4. |9        3
  2      2| 5* |24       5
  1      8| 5. |
          | 6* |144      5
          | 6. |8        2
          | 7* |
          | 7. |9        1
          | 8* |
_____
n:       10       10
_____
```

FIGURE 6.3: Back-to-back stem-and-leaf displays for `cooperation` by `condition`, with the `A` (anonymous-choice) group on the left and the `P` (public-choice) group on the right.

The data from Friendly and Franklin's experiment are in the `Friendly` data set in the `car` package.[3] The data set consists of the variable `correct`, giving the number of words (of the 40) correctly recalled on the final trial of the experiment, and the factor `condition`, with levels `Before`, `Meshed`, and `SFR`. I read the `Friendly` data set in the usual manner (Section 4.2.4), making it the active data set in the R Commander.

Dot plots of the data (Section 5.3.1), in the upper panel of Figure 6.4, suggest that the spreads in the three groups are quite different, and that data in the `Before` group are negatively skewed, at least partly due to a "ceiling effect," with 6 of the 10 subjects recalling 39 or 40 words. This is potentially problematic, because ANOVA assumes normally distributed populations with equal variances.

Using the *Compute New Variable* dialog (described in Section 4.4.2), I add the variable `logit.correct` to the data set—the logit transformation of the proportion of words correctly recalled—computed as `logit(correct/40)`.[4] A dot plot of the transformed data, in the lower panel of Figure 6.4, shows that the spreads in the three groups have been approximately equalized.

With this preliminary work accomplished, I select *Statistics > Means > One-way ANOVA* from the R Commander menus, producing the dialog box in Figure 6.5. The *Groups* variable, `condition`, is preselected because there is only one factor in the data set. I pick `logit.correct` as the *Response Variable*, and check the *Pairwise comparisons* box. Clicking the *OK* button results in the printed output in Figures 6.6 and 6.7, and the graph in Figure 6.8.[5]

As the dot plots in Figure 6.4 suggest, the mean logit of words recalled is greatest in the `Before` condition, nearly as high in the `Meshed` condition, and lowest in the `SFR` condition, while the standard deviations of the logits are similar in the three groups. The differences among the means are just barely statistically significant, with $p = 0.0435$ from the one-way ANOVA.

The pairwise comparisons among the means of the three groups are adjusted for simultaneous inference using Tukey's "honest significant difference" (*HSD*) procedure (Tukey, 1949). The results are shown both as hypothesis tests and as confidence intervals for the pairwise mean differences, with the latter graphed in Figure 6.8. The only pairwise comparison that proves statistically significant is the one between `SFR` and `Before`, $p = 0.047$.[6]

6.1.3   Two-Way and Higher-Way Analysis of Variance

In Section 5.1, I introduced data from an experiment conducted by Adler (1973) on experimenter effects in psychological research. To recapitulate briefly, ostensible "research

---

[3]I'm grateful to Michael Friendly of York University for making the data available.

[4]You may well be unfamiliar with the *logit transformation*, defined as the log of the odds: That is, if p is the proportion correct (i.e., the number correct divided by 40), the *odds* of being correct are $p/(1-p)$ (the proportion correct divided by the proportion incorrect), and the log-odds or logit is $\log[p/(1-p)]$. The logit tends to improve the behavior of proportions that get close to 0 or 1. When, as here, some values of p are *equal* to 0 or 1, the logit is undefined. The `logit` function (from the `car` package) used in computing `logit.correct` moves these extreme proportions slightly away from 0 or 1 prior to calculating the logit, as reflected in a warning printed in the R Commander *Messages* pane.

More generally, don't be concerned about the details of the logit transformation. The object is simply to make the data conform more closely to the assumptions of one-way analysis of variance.

[5]The observant reader working along with the examples in this chapter will notice that, in addition to printed and graphical output, the *One-Way Analysis of Variance* dialog also produces a *statistical model*, named `AnovaModel.1`, which becomes the *active model* in the R Commander. This is true as well of the *Multi-Way Analysis of Variance* dialog discussed in the next section. The active model can then be manipulated via the *Models* menu. The treatment of statistical models in the R Commander is the subject of Chapter 7.

[6]Because Friendly and Franklin (1980) expected higher rates of recall in the `Before` and `Meshed` condition than in the `SFR` condition, it would arguably be legitimate to halve the p-values for the corresponding pairwise comparisons.

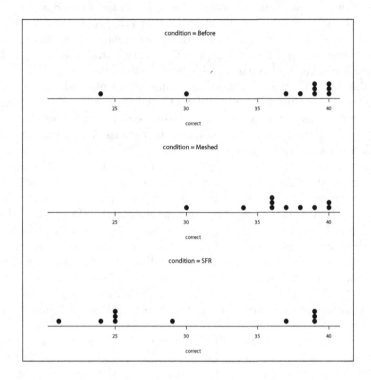

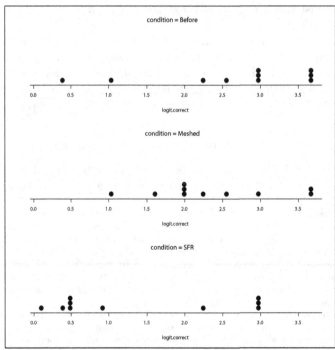

FIGURE 6.4: Dot plots by experimental condition of number (top) and logit (bottom) of words recalled correctly in Friendly and Franklin's memory experiment.

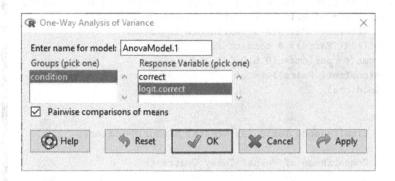

FIGURE 6.5: *One-Way Analysis of Variance* dialog.

```
> AnovaModel.1 <- aov(logit.correct ∼ condition, data=Friendly)

> summary(AnovaModel.1)
            Df Sum Sq Mean Sq F value Pr(>F)
condition    2   8.22   4.110   3.528 0.0435 *
Residuals   27  31.46   1.165
---
Signif. codes:  0 '***' 0.001 '**' 0.01 '*' 0.05 '.' 0.1 ' ' 1

> with(Friendly, numSummary(logit.correct, groups=condition,
+     statistics=c("mean", "sd")))
           mean        sd data:n
Before 2.611010 1.1205637     10
Meshed 2.370046 0.8537308     10
SFR    1.399897 1.2290825     10
```

FIGURE 6.6: One-way ANOVA output for Friendly and Franklin's memory experiment, with the response variable as the logit of words correctly recalled: ANOVA table, means, standard deviations, and counts.

```
> local({
+   .Pairs <- glht(AnovaModel.1, linfct = mcp(condition = "Tukey"))
+   print(summary(.Pairs)) # pairwise tests
+   print(confint(.Pairs)) # confidence intervals
+   print(cld(.Pairs)) # compact letter display
+   old.oma <- par(oma=c(0,5,0,0))
+   plot(confint(.Pairs))
+   par(old.oma)
+ })


        Simultaneous Tests for General Linear Hypotheses

Multiple Comparisons of Means: Tukey Contrasts


Fit: aov(formula = logit.correct ~ condition, data = Friendly)

Linear Hypotheses:
                   Estimate Std. Error t value Pr(>|t|)
Meshed - Before == 0  -0.2410     0.4827  -0.499    0.872
SFR - Before == 0     -1.2111     0.4827  -2.509    0.047 *
SFR - Meshed == 0     -0.9701     0.4827  -2.010    0.129
---
Signif. codes:  0 '***' 0.001 '**' 0.01 '*' 0.05 '.' 0.1 ' ' 1
(Adjusted p values reported -- single-step method)


        Simultaneous Confidence Intervals

Multiple Comparisons of Means: Tukey Contrasts


Fit: aov(formula = logit.correct ~ condition, data = Friendly)

Quantile = 2.4803
95% family-wise confidence level


Linear Hypotheses:
                   Estimate lwr       upr
Meshed - Before == 0 -0.24096 -1.43821  0.95628
SFR - Before == 0    -1.21111 -2.40836 -0.01386
SFR - Meshed == 0    -0.97015 -2.16740  0.22710


Before Meshed    SFR
   "b"    "ab"    "a"
```

FIGURE 6.7: One-way ANOVA output for Friendly and Franklin's memory experiment: Pairwise comparisons of group means.

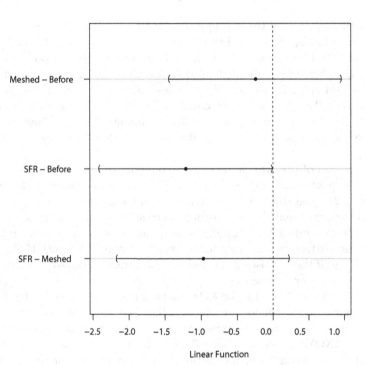

FIGURE 6.8: Simultaneous confidence intervals for pairwise comparisons of group means from Friendly and Franklin's memory experiment, with the response variable as the logit of words correctly recalled.

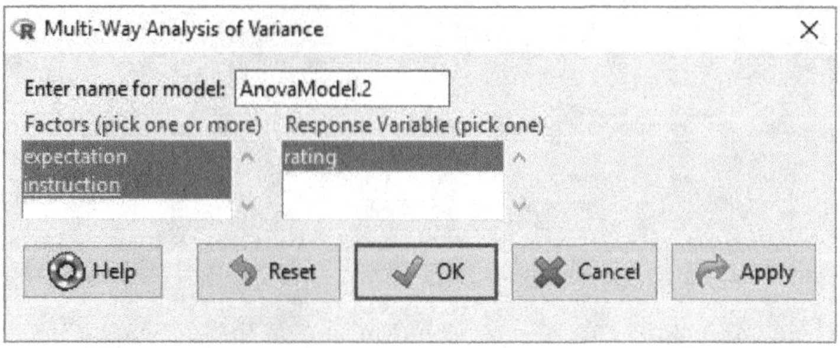

FIGURE 6.9: *Multi-Way Analysis of Variance* dialog box.

assistants" (the actual subjects of the study) were assigned randomly to three experimental conditions, and were variously instructed to collect "good data," "scientific data," or given no such instruction. The research assistants showed photographs to respondents who were asked to rate the apparent "successfulness" of the individuals in the photos, and the assistants were also told, at random, to expect either low or high ratings. The data from the experiment are in the `Adler` data set in the `car` package, which I now read into the R Commander. The data set contains the experimentally manipulated factors `instruction` and `expectation`, and the numeric response variable `rating`, with the average rating obtained by each assistant.

Numerical summaries of the average ratings by combinations of levels of the two factors `instruction` and `expectation` appear in Figure 5.5 (on page 85), and the means are graphed in Figure 5.24 (page 108). The pattern of means suggests an interaction between `instruction` and `expectation`: When assistants were asked to collect "good" data, those with a high expectation produced ratings higher in apparent successfulness than those with low expectations; this pattern was reversed for those told to collect "scientific" data or given no such instruction—as if these assistants leaned over backwards to avoid bias, consequently producing a bias in the reverse direction.

To compute a *two-way ANOVA* for the `Adler` data, I select *Statistics > Means > Multi-way ANOVA*, bringing up the dialog in Figure 6.9.[7] The *Response Variable* `rating` is preselected; I *Ctrl*-click on the *Factors* `expectation` and `instruction`, and click *OK*, obtaining the two-way ANOVA output in Figure 6.10, with an ANOVA table,[8] and cell means, standard deviations, and counts.[9] The interaction between `expectation` and `instruction` proves to be highly statistically significant, $p = 2.6 \times 10^{-6}$.

---

[7]The same dialog can be employed for *higher-way analysis of variance* by selecting more than two factors.

[8]A technical note: Because the `Adler` data are *unbalanced*, with unequal cell counts, there is more than one way to perform the analysis of variance, and the *Multi-Way Analysis of Variance* dialog produces so-called "Type II" tests. Alternative tests are available via *Models > Hypothesis tests > ANOVA table*. For a discussion of this point, see Section 7.7.

[9]The tables of means and standard deviations duplicate those shown in Figure 5.5 on page 85.

```
> AnovaModel.2 <- lm(rating ~ expectation*instruction, data=Adler,
+   contrasts=list(expectation ="contr.Sum", instruction ="contr.Sum"))

> Anova(AnovaModel.2)
Anova Table (Type II tests)

Response: rating
                         Sum Sq Df F value    Pr(>F)
expectation               123.0  1  0.7721    0.3819
instruction               295.6  2  0.9279    0.3991
expectation:instruction  4729.4  2 14.8445 2.633e-06 ***
Residuals               14496.2 91
---
Signif. codes:  0 '***' 0.001 '**' 0.01 '*' 0.05 '.' 0.1 ' ' 1

> with(Adler, (tapply(rating, list(expectation, instruction), mean,
+   na.rm=TRUE))) # means
          GOOD SCIENTIFIC  NONE
HIGH   4.066667  -6.944444 -10.0
LOW  -18.058824   1.923077  -3.5

> with(Adler, (tapply(rating, list(expectation, instruction), sd,
+   na.rm=TRUE))) # std. deviations
        GOOD SCIENTIFIC     NONE
HIGH 16.64961   8.446874 15.63756
LOW  10.57397  11.715101 11.62781

> xtabs(~ expectation + instruction, data=Adler) # counts
            instruction
expectation GOOD SCIENTIFIC NONE
       HIGH   15        18   16
       LOW    17        13   18
```

FIGURE 6.10: Two-way ANOVA output for the `Adler` data: ANOVA table, and cell means, standard deviations, and counts.

6.2   Tests on Proportions

The *Statistics > Proportions* menu (see Figure A.4 on page 202) includes items for single-sample and two-sample tests. Both dialogs are simple and generally similar to the corresponding dialogs for means.

I'll illustrate a single-sample test for a proportion using the **Chile** data set in the **car** package (introduced in Section 5.3.2). The data are from a poll of eligible voters in Chile, conducted about six months prior to the 1988 plebiscite to determine whether the country would continue under military rule or initiate a transition to electoral democracy. A "yes" vote represents support for the continuation of Augusto Pinochet's military government, while a "no" vote represents support for a return to democracy. In the event, of course, the "no" side prevailed in the plebiscite, garnering 56% of the vote.

Of the 2700 voters who were interviewed for the poll, 889 said that they were planning to vote "no" (coded N in the **Chile** data set) and 868 said that they were planning to vote "yes" (coded Y); the remainder of the respondents said that they were undecided (U, 588), were planning to abstain (A, 187), or didn't answer the question (NA, 168). After reading the **Chile** data set in the usual manner, I begin my analysis by recoding **vote** into a two-level factor, retaining the variable name **vote**, and employing the recode directives "Y" = "yes", "N" = "no", and else = NA (see Section 4.4.1).

Choosing *Statistics > Proportions > Single-sample proportion test* produces the dialog in Figure 6.11. I select **vote** in the *Data* tab and leave the selections in the *Options* tab at their defaults. In particular, the default null hypothesis that the population proportion is 0.5 against the default two-sided alternative makes sense in the context of the pre-plebiscite poll.

Clicking *OK* generates a test, shown at the top of Figure 6.12, based on the normal approximation to the binomial distribution. For comparison, I've also shown (at the bottom of Figure 6.12) the output from the exact binomial test, obtained by selecting the corresponding radio button in the *Single-Sample Proportions Test* dialog. The proportion reported is for the **no** level of the two-level factor **vote**—the level that is alphabetically first. Because of the large sample size and the sample proportion close to 0.5, the normal approximation to the binomial is very accurate, even without using a continuity correction (which is also available in the dialog). With the sample proportion $\widehat{p}_{no} = 0.506$, the null hypothesis of a dead heat can't be rejected, *p*-value = 0.63 (by the exact test). Close polling results like this motivated the "no" campaign in the Chilean plebiscite to greater unity and effort.

The **prop.test** function reports a chi-square test statistic (labelled **X-squared**) on one degree of freedom for the test based on the normal approximation to the binomial. It is more common to use the normally distributed test statistic

$$z = \frac{\widehat{p} - p_0}{\sqrt{p_0(1 - p_0)/n}}$$

where $\widehat{p}$ is the sample proportion, p_0 is the hypothesized population proportion (0.5 in the example), and n is the sample size. The relationship between the two test statistics is very simple, $X^2 = z^2$, and they produce identical *p*-values (because a chi-square random variable on one *df* is the square of a standard-normal random variable).

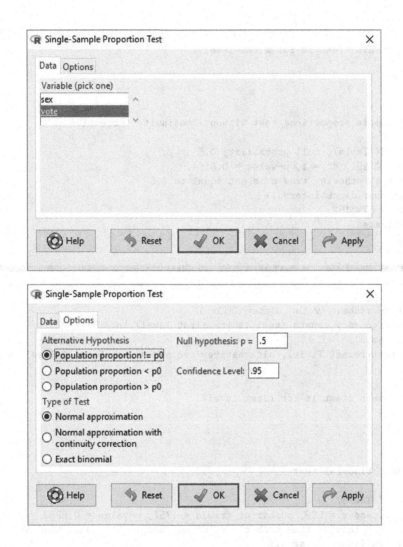

FIGURE 6.11: *Single-Sample Proportions Test* dialog, with *Data* and *Options* tabs.

```
> local({
+    .Table <- xtabs(~ vote , data= Chile )
+    cat("\nFrequency counts (test is for first level):\n")
+    print(.Table)
+    prop.test(rbind(.Table), alternative='two.sided', p=.5, conf.level=.95,
+    correct=FALSE)
+ })

Frequency counts (test is for first level):
vote
 no yes
889 868

        1-sample proportions test without continuity correction

data:  rbind(.Table), null probability 0.5
X-squared = 0.251, df = 1, p-value = 0.6164
alternative hypothesis: true p is not equal to 0.5
95 percent confidence interval:
 0.4826109 0.5293152
sample estimates:
        p
0.5059761
```

```
> local({
+    .Table <- xtabs(~ vote , data= Chile )
+    cat("\nFrequency counts (test is for first level):\n")
+    print(.Table)
+    binom.test(rbind(.Table), alternative='two.sided', p=.5, conf.level=.95)
+ })

Frequency counts (test is for first level):
vote
 no yes
889 868

        Exact binomial test

data:  rbind(.Table)
number of successes = 889, number of trials = 1757, p-value = 0.6333
alternative hypothesis: true probability of success is not equal to 0.5
95 percent confidence interval:
 0.4823204 0.5296118
sample estimates:
probability of success
           0.5059761
```

FIGURE 6.12: Single-sample proportion tests: normal approximation to the binomial (top) and exact binomial test (bottom).

6.3   Tests on Variances

Tests on variances are generally not recommended as preliminaries to tests on means but may be of interest in their own right. The *Statistics > Variances* menu (shown in Figure A.4 on page 202) includes the *F-test* for the difference between two variances, *Bartlett's test* for differences among several variances, and *Levene's test* for differences among several variances. Of these tests, Levene's test is most robust with respect to departures from normality, and so I'll use it to illustrate, returning to Friendly and Franklin's memory data (introduced in Section 6.1.2 on one-way ANOVA): I select `Friendly` via the *Data set* button in the R Commander toolbar to make it the active data set.

The *Levene's Test* dialog is depicted in Figure 6.13. Because there is only one factor in the `Friendly` data set, `condition` is preselected in the dialog. More generally, several factors can be selected, in which case groups are defined by combinations of levels of the factors. I select `correct`—the number of words correctly remembered—as the response; recall that I computed the logit transformation of the proportion correct partly to equalize the variances in the different conditions. I leave the *Center* specification at its default, to use the *median*, the more robust of the two choices. These selections lead to the output in Figure 6.14, showing that the differences in variation among the three conditions are not quite statistically significant ($p = 0.078$).

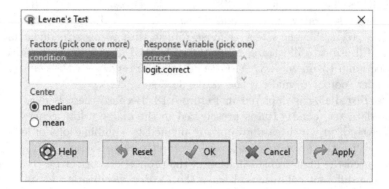

FIGURE 6.13: The *Levene's Test* dialog box to test for differences among several variances.

```
> leveneTest(correct ~ condition, data=Friendly, center="median")
Levene's Test for Homogeneity of Variance (center = "median")
      Df F value  Pr(>F)
group  2  2.8078 0.07801 .
      27
---
Signif. codes:  0 '***' 0.001 '**' 0.01 '*' 0.05 '.' 0.1 ' ' 1
```

FIGURE 6.14: Levene's test for differences in variation of number of words recalled among the three conditions in Friendly and Franklin's memory experiment.

6.4   Nonparametric Tests

The *Statistics > Nonparametric tests* menu (see Figure A.4 on page 202) contains items for several common nonparametric tests—that is, tests that don't make distributional assumptions about the populations sampled. These tests include single-sample, two-sample, and paired-samples *Wilcoxon signed-rank tests* (also known as *Mann–Whitney tests*), which are nonparametric alternatives to single-sample, two-sample, and paired-samples t tests for means; the *Kruskal–Wallis test*, which is a nonparametric alternative to one-way ANOVA (and is a generalization of the two-sample Wilcoxon test); and the *Friedman rank-sum test*, which is an extension of a matched-pairs test to more than two matched values, often used for repeated measures on individuals.

All of the nonparametric test dialogs in the R Commander are very simple. I'll use the Kruskal–Wallis test for Friendly and Franklin's memory data as an example. A one-way ANOVA for the logit of the proportion of words correctly recalled by each subject in the experiment was performed in Section 6.1.2. Choosing *Statistics > Nonparametric tests > Kruskal-Wallis test* produces the dialog box in Figure 6.15. The factor condition is preselected to define *Groups*, and I pick correct as the response variable, obtaining the output in Figure 6.16. The Kruskal–Wallis test for differences among the conditions in the Friendly and Franklin memory experiment is not quite statistically significant ($p = 0.075$).

In contradistinction to the parametric one-way analysis of variance, I would get *exactly the same* Kruskal–Wallis test whether I use correct or logit.correct as the response variable (try it!), because the Kruskal–Wallis test is based on the *ranks* of the response values and not directly on the values themselves.[10]

---

[10]The group medians, also reported by the Kruskal–Wallis dialog, differ, however, depending on which response variable is used, although even here there is a fundamental invariance: The median of the logit-transformed data is the logit of the median (with slight slippage due to interpolation when—as in this example—a median is computed for an even number of values).

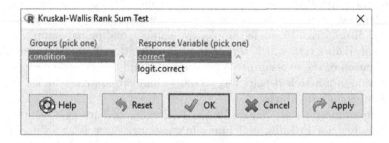

FIGURE 6.15: *Kruskal-Wallis Rank Sum Test* dialog box.

```
> with(Friendly, tapply(correct, condition, median, na.rm=TRUE))
Before Meshed    SFR
  39.0   36.5   27.0

> kruskal.test(correct ~ condition, data=Friendly)

        Kruskal-Wallis rank sum test

data:  correct by condition
Kruskal-Wallis chi-squared = 5.1817, df = 2, p-value = 0.07496
```

FIGURE 6.16: Kruskal–Wallis test for differences among conditions in Friendly and Franklin's memory experiment.

6.5 Other Simple Tests*

The *Statistics > Summaries* menu contains menu items for two simple statistical tests: The Shapiro–Wilk test of normality, and tests for Pearson product-moment and rank correlation coefficients. These tests are for numeric variables. I'll illustrate using the `Prestige` data set, introduced in Section 4.2.3 and available in the **car** package, reading the data set as usual via *Data > Data in packages > Read data from an attached package.*

The distributional displays in Figure 5.10 (on page 92) for `education` in the `Prestige` data set suggest that the variable isn't normally distributed: The histogram and density plot for the data appear to have multiple modes, and the normal quantile-comparison plot shows shorter tails than a normal distribution. Selecting *Statistics > Summaries > Shapiro-Wilk test of normality* brings up the dialog box in Figure 6.17; selecting `education` and clicking *OK* produces the output in Figure 6.18. The departure from normality is highly statistically significant ($p = 0.00068$).

The scatterplot for `prestige` versus `education` in the `Prestige` data set, shown in Figure 5.16 (page 99) suggests a monotone (strictly increasing) but nonlinear relationship between the two variables. Selecting *Statistics > Summaries > Correlation test* from the R Commander menus leads to the dialog box in Figure 6.19. I select `education` and `income` in the *Variables* list box, and, because the relationship between the two variables is apparently nonlinear, I'll examine a rank-order correlation rather than the Pearson product-moment correlation between the variables. There are ties in the data, and so I select *Kendall's tau* in preference to *Spearman's rank-order* correlation. I anticipate a positive correlation between `education` and `income`, as reflected in the choice of *Alternative Hypothesis.* The output produced by clicking *OK*, shown in Figure 6.20, indicates that the positive ordinal relationship between the two variables is very highly statistically significant, with $p = 5.5 \times 10^{-10}$, effectively 0; the estimated Kendall correlation is $\hat{\tau} = 0.41$.

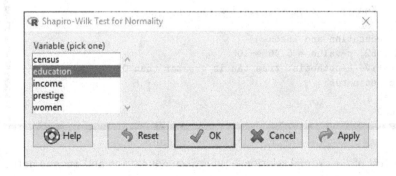

FIGURE 6.17: *Shapiro-Wilk Test for Normality* dialog box.

```
> with(Prestige, shapiro.test(education))

        Shapiro-Wilk normality test

data:  education
W = 0.94958, p-value = 0.0006773
```

FIGURE 6.18: Output for the Shapiro–Wilk normality test of education in the Prestige data set.

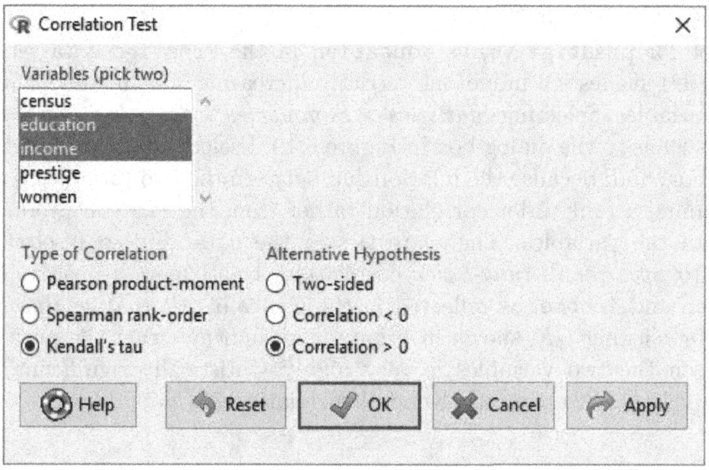

FIGURE 6.19: *Correlation Test* dialog box.

```
> with(Prestige, cor.test(education, income, alternative="greater",
+   method="kendall"))

        Kendall's rank correlation tau

data:  education and income
z = 6.0952, p-value = 5.465e-10
alternative hypothesis: true tau is greater than 0
sample estimates:
     tau
0.409559
```

FIGURE 6.20: Test of the ordinal relationship between education and income in the Prestige data set.

7

Fitting Linear and Generalized Linear Models

Beyond basic statistics, *regression models* of various kinds are at the heart of applied statistical methods. Regression models trace how the distribution of a response (or "dependent") variable—or some key characteristics of that distribution, such as its mean—is related to the values of one or more explanatory ("independent") variables. Least-squares linear regression is typically introduced in a basic statistics course, while a more general consideration of linear statistical models for normally distributed responses and generalized linear models for non-normally distributed responses is typically the subject of a second course in applied statistics (see, e.g., Fox, 2016, or Weisberg, 2014).

This chapter explains how to fit linear and generalized linear regression models in the R Commander, and how to perform additional computations on regression models once they have been fit to data. By treating statistical models as *objects* subject to further computation, R encourages a style of statistical modeling in which the data analyst engages in a back-and-forth conversation with the data. As I will demonstrate, the R Commander takes advantage of this orientation.

7.1 Linear Regression Models

As mentioned, linear least-squares regression is typically taken up in a basic statistics course. The *normal linear regression model* is written

$$
\begin{aligned}
y_i &= \beta_0 + \beta_1 x_{1i} + \beta_2 x_{2i} + \cdots + \beta_k x_{ki} + \varepsilon_i \\
&= E(y_i) + \varepsilon_i
\end{aligned}
\tag{7.1}
$$

where y_i is the value of the response variable for the ith of n independently sampled observations; $x_{1i}, x_{2i}, \ldots, x_{ki}$ are the values of k explanatory variables; and the errors ε_i are normally and independently distributed with 0 means and constant variance, $\varepsilon_i \sim \text{NID}(0, \sigma_\varepsilon^2)$. Both y and the xs are numeric variables, and the model assumes that the average value $E(y)$ of y is a linear function—that is, a simple weighted sum—of the xs.[1] If there is just one x (i.e., if $k = 1$), then Equation 7.1 is called the *linear simple regression model*; if there are more than one x ($k \geq 2$), then it is called the *linear multiple regression model*.

The normal linear model is optimally estimated by the *method of least squares*, producing the *fitted model*

$$
\begin{aligned}
y_i &= b_0 + b_1 x_{1i} + b_2 x_{2i} + \cdots + b_k x_{ki} + e_i \\
&= \widehat{y}_i + e_i
\end{aligned}
$$

where $\widehat{y}_i$ is the *fitted value* and e_i the *residual* for observation i. The least-squares criterion

---

[1] As explained in Section 7.2, Equation 7.1 also serves for more general linear models, where (some of) the xs may not be numeric explanatory variables but rather dummy regressors representing factors, interaction regressors, polynomial regressors, and so on.

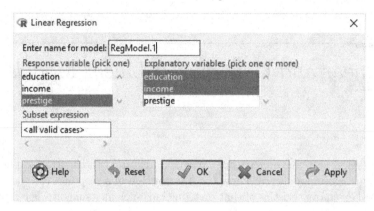

FIGURE 7.1: The *Linear Regression* dialog for Duncan's occupational prestige data.

selects the values of the bs that minimize the sum of squared residuals, $\sum e_i^2$. The least-squares regression coefficients are easily computed, and, in addition to having desirable statistical properties under the model (such as efficiency and unbias), statistical inference based on the least-squares estimates is very simple (see, e.g., the references given at the beginning of the chapter).

The simplest way to fit a linear regression model in the R Commander is by the *Linear Regression* dialog. To illustrate, I'll use Duncan's occupational prestige data (introduced in Chapter 4). Duncan's data set resides in the **car** package, and so I can read the data into the R Commander via *Data > Data in packages > Read data from an attached package* (see Section 4.2.4). Then selecting *Statistics > Fit models > Linear regression* produces the dialog in Figure 7.1. To complete the dialog, I click on **prestige** in the *Response variable* list, and *Ctrl*-click on **education** and **income** in the *Explanatory variables* list. Finally, pressing the *OK* button produces the output shown in Figure 7.2.

The commands generated by the *Linear Regression* dialog use the **lm** (linear model) function in R to fit the model, creating **RegModel.1**, and then summarize the model to produce printed output. The summary output includes information about the distribution of the residuals; coefficient estimates, their standard errors, t statistics for testing the null hypothesis that each population regression coefficient is 0, and the two-sided p-values for these tests; the standard deviation of the residuals ("residual standard error") and residual degrees of freedom; the squared multiple correlation, R^2, for the model and R^2 adjusted for degrees of freedom; and the omnibus F test for the hypothesis that all population slope coefficients (here the coefficients of **education** and **income**) are 0 (H_0: $\beta_1 = \beta_2 = 0$, for the example).

This is more or less standard least-squares regression output, similar to printed output produced by almost all statistical packages. What is unusual is that in addition to the printout in Figure 7.2, the R Commander creates and retains a *linear model object* on which I can perform further computations, as illustrated later in this chapter.

The *Model* button in the R Commander toolbar now reads *RegModel.1*, rather than *<No active model>*, as it did at the beginning the session. Just as you can choose among data sets residing in memory (if there are more than one) by pressing the *Data set* button in the toolbar, you can similarly choose among statistical models (if there are more than one) by pressing the *Model* button. Equivalently, you can pick *Models > Select active model* from the R Commander menus. Moreover, the R Commander takes care of coordinating data sets and models, by associating each statistical model with the data set to which it is fit.

```
> RegModel.1 <- lm(prestige~education+income, data=Duncan)

> summary(RegModel.1)

Call:
lm(formula = prestige ~ education + income, data = Duncan)

Residuals:
    Min      1Q  Median      3Q     Max
-29.538  -6.417   0.655   6.605  34.641

Coefficients:
            Estimate Std. Error t value Pr(>|t|)
(Intercept) -6.06466    4.27194  -1.420    0.163
education    0.54583    0.09825   5.555 1.73e-06 ***
income       0.59873    0.11967   5.003 1.05e-05 ***
---
Signif. codes:  0 '***' 0.001 '**' 0.01 '*' 0.05 '.' 0.1 ' ' 1

Residual standard error: 13.37 on 42 degrees of freedom
Multiple R-squared:  0.8282,    Adjusted R-squared:    0.82
F-statistic: 101.2 on 2 and 42 DF,  p-value: < 2.2e-16
```

FIGURE 7.2: Output from Duncan's regression of occupational `prestige` on `income` and `education`, produced by the *Linear Regression* dialog.

Consequently, selecting a statistical model makes the data set to which it was fit the active data set, if that isn't already the case.

The variable lists in the *Linear Regression* dialog in Figure 7.1 include only numeric variables. For example, the factor `type` (type of occupation) in Duncan's data set, with levels `"bc"` (blue-collar), `"wc"` (white-collar), and `"prof"` (professional, technical, or managerial), doesn't appear in either variable list. Moreover, the explanatory variables that are selected enter the model linearly and additively. The *Linear Model* dialog, described in the next section, is capable of fitting a much wider variety of regression models.

In completing the *Linear Regression* dialog in Figure 7.1, I left the name of the model at its default, `RegModel.1`. The R Commander generates unique model names automatically during a session, each time incrementing the model number (here 1).

I also left the *Subset expression* at its default, `<all valid cases>`. Had I instead entered `type == "bc"`,[2] for example, the regression model would have been fit only to blue-collar occupations. As in this example, the subset expression can be a logical expression, returning the value `TRUE` or `FALSE` for each case (see Section 4.4.2), a vector of case indices to include,[3] or a negative vector of case indices to *exclude*. For example, `1:25` would include the first 25 occupations, while `-c(6, 16)` would exclude occupations 6 and 16.[4] All of the statistical modeling dialogs in the R Commander allow subsets of cases to be specified in this manner.

---

[2]Recall that you must use the double equals sign (`==`) to test for equality; see Table 4.4 (page 71).

[3]A *vector* is a one-dimensional array, here of numbers.

[4]The *sequence operator* `:` creates an integer (whole-number) sequence, so `1:25` generates the integers 1 through 25. The c function *combines* its arguments into a vector, so `-c(6, 16)` creates a two-element vector containing the numbers −6 and −16.

7.2   Linear Models with Factors*

Like the *Linear Regression* dialog described in the preceding section, the *Linear Model* dialog can fit additive linear regression models, but it is much more flexible: The *Linear Model* dialog accommodates transformations of the response and explanatory variables, factors as well as numeric explanatory variables on the right-hand-side of the regression model, nonlinear functions of explanatory variables expressed as polynomials and regression splines, and interactions among explanatory variables. All this is accomplished by allowing the user to specify the model as an R *linear-model formula*. Linear-model formulas in R are inherited from the S programming language (Chambers and Hastie, 1992), and are a version of notation for expressing linear models originally introduced by Wilkinson and Rogers (1973).

7.2.1   Linear-Model Formulas

An R linear-model formula is of the general form `response-variable ~ linear-predictor`. The tilde (~) in a linear-model formula can be read as "is regressed on." Thus, in this general form, the response variable is regressed on a linear predictor comprising the *terms* in the right-hand side of the model.

The left-hand side of the model formula, `response-variable`, is an R expression that evaluates to the numeric response variable in the model, and is usually simply the *name* of the response variable—for example, `prestige` in Duncan's regression. You can, however, transform the response variable directly in the model formula (e.g., `log10(income)`) or compute the response as a more complex arithmetic expression (e.g., `log(investment.income + hourly.wage.rate*hours.worked)`.[5]

The formulation of the linear predictor on the right-hand side of a model formula is more complex. What are normally arithmetic operators (`+`, `-`, `*`, `/`, and `^`) in R expressions have special meanings in a model formula, as do the operators `:` (colon) and `%in%`. The numeral `1` (one) may be used to represent the regression constant (i.e., the intercept) in a model formula; this is usually unnecessary, however, because an intercept is included by default. A period (`.`) represents all of the variables in the data set with the exception of the response. Parentheses may be used for grouping, much as in an arithmetic expression.

In the large majority of cases, you'll be able to formulate a model using only the operators `+` (interpreted as "and") and `*` (interpreted as "crossed with"), and so I'll emphasize these operators here. The meaning of these and the other model-formula operators are summarized and illustrated in Table 7.1. Especially on first reading, feel free to ignore everything in the table except `+`, `:`, and `*` (and `:` is rarely used directly).

A final formula subtlety: As I've explained, the arithmetic operators take on special meanings on the right-hand side of a linear-model formula. A consequence is that you can't use these operators directly for arithmetic. For example, fitting the model `savings ~ wages + interest + dividends` estimates a *separate* regression coefficient for each of `wages`, `interest`, and `dividends`. Suppose, however, that you want to estimate a *single* coefficient for the sum of these variables—in effect, setting the three coefficients equal to each other. The solution is to "protect" the `+` operator inside a call to the `I` (*identity* or *inhibit*) function, which simply returns its argument unchanged:[6] `savings ~ I(wages + interest + dividends)`. This formula works as desired because arithmetic operators like `+` have their usual meaning *within* a function call on the right-hand side of the formula—implying,

---

[5]See Section 4.4.2, and in particular Table 4.4 (page 71), for information on R expressions.

[6]The *arguments* of an R function are the values given in parentheses when the function is called; if there is more than one argument, they are separated by commas.

TABLE 7.1: Operators and other symbols used on the right-hand side of R linear-model formulas.

Operator	Meaning	Example	Interpretation
+	and	x1 + x2	x1 and x2
:	interaction	x1:x2	interaction of x1 and x2
*	crossing	x1*x2	x1 crossed with x2 (i.e., x1 + x2 + x1:x2)
-	remove	x1 - 1	regression through the origin (for numeric x1)
^k	cross to order k	(x1 + x2 + x3)^2	same as x1*x2 + x1*x3 + x2*x3
%in%	nesting	province %in% country	province nested in country
/	nesting	country/province	same as country + province %in% country

Symbol	Meaning	Example	Interpretation
1	intercept	x1 - 1	suppress the intercept
.	everything but the response	y ~ .	regress y on everything else
( )	grouping	x1*(x2 + x3)	same as x1*x2 + x1*x3

The symbols x1, x2, and x3 represent explanatory variables and could be either numeric or factors.

incidentally, that `savings ~ log10(wages + interest + dividends)` also works as intended, estimating a single coefficient for the log base 10 of the sum of `wages`, `interest`, and `dividends`.

7.2.2 The Principle of Marginality

Introduced by Nelder (1977), the *principle of marginality* is a rule for formulating and interpreting linear (and similar) statistical models. According to the principle of marginality, if an *interaction*, say x1:x2, is included in a linear model, then so should the *main effects*, x1 and x2, that are *marginal* to—that is *lower-order relatives* of—the interaction. Similarly, the *lower-order interactions* x1:x2, x1:x3, and x2:x3 are marginal to the *three-way interaction* x1:x2:x3. The regression constant (1 in an R model formula) is marginal to every other term in the model.[7]

It is in most circumstances difficult in R to formulate models that violate the principle of marginality, and trying to do so can produce unintended results. For example, although it may appear that the model y ~ f*x - x - 1, where f is a factor and x is a numeric explanatory variable,[8] violates the principle of marginality by removing the regression constant and x slope, the model that R actually fits includes a separate intercept and slope for each level of the factor f. Thus, the model y ~ f*x - x - 1 is equivalent to (i.e., an alternative parametrization of) y ~ f*x. It is almost always best to stay away from such unusual model formulas.

7.2.3 Examples Using the Canadian Occupational Prestige Data

For concreteness, I'll formulate several linear models for the Canadian occupational prestige data (introduced in Section 4.2.3 and described in Table 4.2 on page 61), regressing pres-

---

[7]As formulated by Nelder (1977), the principle of marginality is deeper and more general than this characterization, but thinking of the principle in these simplified terms will do for our purposes.

[8]See later in this section for an explanation of how factors are handled in linear-model formulas.

tige on income, education, women (gender composition), and type (type of occupation). The last variable is a factor (categorical variable) and so it cannot enter into the linear model directly. When a factor is included in a linear-model formula, R generates *contrasts* to represent the factor—one fewer than the number of levels of the factor. I'll explain how this works in greater detail in Section 7.2.4, but the default in the R Commander (and R more generally) is to use 0/1 *dummy-variable regressors*, also called *indicator variables*.

A version of the Canadian occupational prestige data resides in the data frame Prestige in the **car** package,[9] and it's convenient to read the data into the R Commander from this source via *Data > Data in packages > Read data from an attached package*. Prestige replaces Duncan as the active data set.

Recall that 4 of the 102 occupations in the Prestige data set have missing values (NA) for occupational type. Because I will fit several regression models to the Prestige data, not all of which include type, I begin by filtering the data set for missing values, selecting *Data > Active data set > Remove cases with missing data* (as described in Section 4.5.2).

Moreover, the default alphabetical ordering of the levels of type—"bc", "prof", "wc"— is not the natural ordering, and so I also reorder the levels of this factor via *Data > Manage variables in active data set > Reorder factor levels* to "bc", "wc", "prof" (see Section 3.4). This last step isn't strictly necessary, but it makes the data analysis easier to follow.

I first fit an additive dummy regression to the Canadian prestige data, employing the model formula prestige ~ income + education + women + type. To do so, I select *Statistics > Fit models > Linear model* from the R Commander menus, producing the dialog box in Figure 7.3. The automatically supplied model name is LinearModel.2, reflecting the fact that I have already fit a statistical model in the session, RegModel.1 (in Section 7.1).

Most of the structure of the *Linear Model* dialog is common to statistical modeling dialogs in the R Commander. If the response text box to the left of the ~ in the model formula is empty, double-clicking on a variable name in the variable list box enters the name into the response box; thereafter, double-clicking on variable names enters the names into the right-hand side of the model formula, separated by +s (if no operator appears at the end of the partially completed formula). You can enter parentheses and operators like + and * into the formula using the toolbar in the dialog box.[10] You can also type directly into the model-formula text boxes. In Figure 7.3, I simply double-clicked successively on prestige, education, income, women, and type.[11] Clicking *OK* produces the output shown in Figure 7.4.

I already explained the general format of linear-model summary output in R. What's new in Figure 7.4 is the way in which the factor type is handled in the linear model: Two dummy-variable regressors are automatically created for the three-level factor type. The first dummy regressor, labelled type[T.wc] in the output, is coded 1 when type is "wc" and 0 otherwise; the second dummy regressor, type[T.prof], is coded 1 when type is "prof" and 0 otherwise. The first level of type—"bc"—is therefore selected as the *reference* or *baseline level*, coded 0 for both dummy regressors.[12]

Consequently, the intercept in the linear-model output is the intercept for the "bc" reference level of type, and the coefficients for the other levels give differences in the intercepts

---

[9]See Table 4.2 (on page 61) for an explanation of the variables in the Prestige data set.

[10]The second toolbar may be used to enter regression-spline and polynomial terms into the model. I'll describe this feature in Section 7.3. If you're unfamiliar with regression splines or polynomials, simply ignore this toolbar.

[11]The *Linear Model* dialog also includes a box for subsetting the data, and a drop-down variable list for selecting a weight variable for *weighted-least-squares* (as opposed to *ordinary-least-squares*) regression. Neither subsetting nor a weight variable is used in this example.

[12]The "T" in the names of the dummy-variable coefficients refers to "treatment contrasts"—a synonym for 0/1 dummy regressors—discussed further in Section 7.2.4.

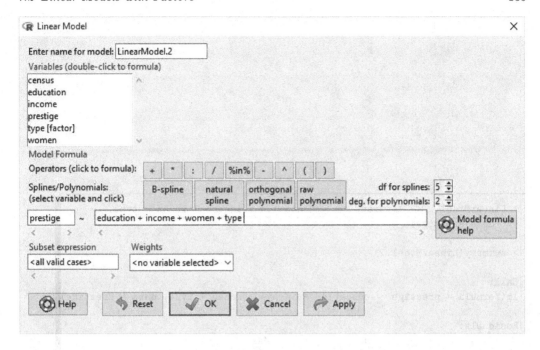

FIGURE 7.3: *Linear Model* dialog completed to fit an additive dummy-variable regression of `prestige` on the numeric explanatory variables `education`, `income`, and `women`, and the factor `type`.

between each of these levels and the reference level. Because the slope coefficients for the numeric explanatory variables `education`, `income`, and `women` in this additive model do not vary by levels of `type`, the dummy-variable coefficients are also interpretable as the average difference between each other level and `"bc"` for *any* fixed values of `education`, `income`, and `women`.

To illustrate a structurally more complex, nonadditive model, I respecify the Canadian occupational prestige regression model to include interactions between `type` and `education` and between `type` and `income`, in the process removing `women` from the model—in the initial regression, the coefficient of `women` is small with a large p-value.[13] The *Linear Model* dialog (not shown) reopens in its previous state, with the model name incremented to `LinearModel.3`. To fit the new model, I modify the formula to read `prestige ~ type*education + type*income`. Clicking *OK* produces the output in Figure 7.5.

With interactions in the model, there are different intercepts and slopes for each level of `type`. The intercept in the output—along with the coefficients for `education` and `income`—pertains to the baseline level `"bc"` of `type`. Other coefficients represent differences between each of the other levels and the baseline level. For example, `type[T.wc]` $= -33.54$ is the difference in intercepts between the `"wc"` and `"bc"` levels of `type`;[14] similarly, the interaction coefficient `type[T.wc]:education` $= 4.291$ is the difference in `education` slopes between the `"wc"` and `"bc"` levels. The complexity of the coefficients makes it difficult to understand

---

[13]That the coefficient of `women` in the additive model is small doesn't imply that `women` and `type` don't interact. My true motive here is to simplify the example.

[14]Recall that a number like `-3.354e+01` is expressed in scientific notation, and is written conventionally as $-3.354 \times 10^1 = -33.54$.

```
> LinearModel.2 <- lm(prestige ~ education + income + women + type,
+    data=Prestige)

> summary(LinearModel.2)

Call:
lm(formula = prestige ~ education + income + women + type, data = Prestige)

Residuals:
     Min      1Q   Median       3Q      Max
-14.7485  -4.4817   0.3119   5.2478  18.4978

Coefficients:
              Estimate Std. Error t value Pr(>|t|)
(Intercept)  -0.8139032  5.3311558  -0.153 0.878994
education     3.6623557  0.6458300   5.671 1.63e-07 ***
income       0.0010428  0.0002623   3.976 0.000139 ***
women        0.0064434  0.0303781   0.212 0.832494
type[T.wc]   -2.9170720  2.6653961  -1.094 0.276626
type[T.prof]  5.9051970  3.9377001   1.500 0.137127
---
Signif. codes:  0 '***' 0.001 '**' 0.01 '*' 0.05 '.' 0.1 ' ' 1

Residual standard error: 7.132 on 92 degrees of freedom
Multiple R-squared:  0.8349,    Adjusted R-squared:  0.826
F-statistic: 93.07 on 5 and 92 DF,  p-value: < 2.2e-16
```

FIGURE 7.4: Output for the linear model prestige ~ income + education + women + type fit to the Prestige data.

```
> LinearModel.3 <- lm(prestige ~ type*education + type*income, data=Prestige)

> summary(LinearModel.3)

Call:
lm(formula = prestige ~ type * education + type * income, data = Prestige)

Residuals:
    Min      1Q  Median      3Q     Max
-13.462  -4.225   1.346   3.826  19.631

Coefficients:
                       Estimate Std. Error t value Pr(>|t|)
(Intercept)           2.276e+00  7.057e+00   0.323   0.7478
type[T.wc]           -3.354e+01  1.765e+01  -1.900   0.0607 .
type[T.prof]          1.535e+01  1.372e+01   1.119   0.2660
education             1.713e+00  9.572e-01   1.790   0.0769 .
income                3.522e-03  5.563e-04   6.332 9.62e-09 ***
type[T.wc]:education   4.291e+00  1.757e+00   2.442   0.0166 *
type[T.prof]:education 1.388e+00  1.289e+00   1.077   0.2844
type[T.wc]:income    -2.072e-03  8.940e-04  -2.318   0.0228 *
type[T.prof]:income  -2.903e-03  5.989e-04  -4.847 5.28e-06 ***
---
Signif. codes:  0 '***' 0.001 '**' 0.01 '*' 0.05 '.' 0.1 ' ' 1


Residual standard error: 6.318 on 89 degrees of freedom
Multiple R-squared:  0.8747,     Adjusted R-squared:  0.8634
F-statistic: 77.64 on 8 and 89 DF,  p-value: < 2.2e-16
```

FIGURE 7.5: Output for the linear model prestige ~ type*education + type*income fit to the Prestige data.

TABLE 7.2: Contrast-regressor codings for `type` generated by `contr.Treatment`, `contr.Sum`, `contr.poly,`, and `contr.Helmert`.

Function	Contrast Names	Levels of type "bc"	"wc"	"prof"
contr.Treatment	type[T.wc]	0	1	0
	type[T.prof]	0	0	1
contr.Sum	type[S.wc]	1	0	-1
	type[S.prof]	0	1	-1
contr.poly	type.L	$-1/\sqrt{2}$	0	$1/\sqrt{2}$
	type.Q	$1/\sqrt{6}$	$-2/\sqrt{6}$	$1/\sqrt{6}$
contr.Helmert	type[H.1]	-1	1	0
	type[H.2]	-1	-1	2

what the model says about the data; Section 7.6 shows how to visualize terms such as interactions in a complex linear model.

7.2.4   Dummy Variables and Other Contrasts for Factors

By default in the R Commander, factors in linear-model formulas are represented by 0/1 dummy-variable regressors generated by the `contr.Treatment` function in the **car** package, picking the first level of a factor as the baseline level.[15] This contrast coding, along with some other choices, is shown in Table 7.2, using the factor `type` in the **Prestige** data set as an example.

The function `contr.Sum` from the **car** generates so-called "sigma-constrained" or "sum-to-zero" contrast regressors, as are used in traditional treatments of analysis of variance.[16] The standard R function `contr.poly` generates orthogonal-polynomial contrasts—in this case, linear and quadratic terms for the three levels of `type`; in the R Commander, `contr.poly` is the default choice for ordered factors. Finally, `contr.Helmert` generates Helmert contrasts, which compare each level to the average of those preceding it.

Selecting *Data > Manage variables in active data set > Define contrasts for a factor* produces the dialog box on the left of Figure 7.6. The factor `type` is preselected in this dialog because it's the only factor in the data set. You can use the radio buttons to choose among treatment, sum-to-zero, Helmert, and polynomial contrasts, or define customized contrasts by selecting *Other*, as I've done here.

Clicking *OK* leads to the sub-dialog shown on the right of Figure 7.6. I change the default contrast names, .1 and .2, to [bc.v.others] and [wc.v.prof], and then fill in the contrast coefficients (i.e., the values of the contrast regressors). This choice produces contrast regressors named `type[bc.v.others]` and `type[wc.v.prof]`, to be used when the factor `type` in the **Prestige** data set appears in a linear-model formula. Contrasts defined directly in this manner must be linearly independent and are simplest to interpret if they obey two additional rules: (1) The coefficients for each contrast should sum to 0, and (2)

---

[15]The function `contr.Treatment` is a modified version of the standard R function `contr.treatment`; `contr.Treatment` generates slightly easier to read names for the dummy variables—for example, `type[T.wc]` rather than `typewc`. Similarly, `contr.Sum` and `contr.Helmert`, discussed below, are modifications of the standard R functions `contr.sum` and `contr.helmert`.

[16]Multi-way analysis of variance in the R Commander, discussed in Section 6.1.3, uses `contr.Sum` for the factors in the ANOVA model.

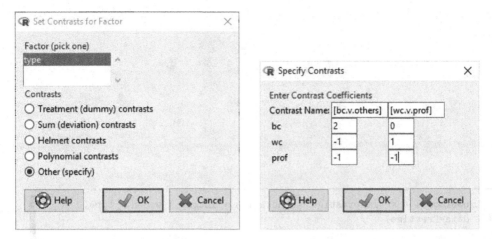

FIGURE 7.6: The *Set Contrasts for Factor* dialog box (left) and the *Specify Contrasts* sub-dialog (right), creating contrasts for the factor `type` in the `Prestige` data set.

each pair of contrasts should be orthogonal (i.e., the products of corresponding coefficients for each pair of contrasts sum to 0).

To see how these contrasts are reflected in the coefficients of the model, I refit the additive regression of `prestige` on `education`, `income`, `women`, and `type`, producing the output in Figure 7.7. The first contrast for `type` estimates the difference between `"bc"` and the average of the other two levels of `type`, holding the other explanatory variables constant, while the second contrast estimates the difference between `"wc"` and `"prof"`. This alternative contrast coding for `type` produces different estimates for the intercept and `type` coefficients from the dummy-regressor coding for `type` in Figure 7.4 (on page 136), but the two models have the same fit to the data (e.g., $R^2 = 0.8349$).[17]

---

[17]Reader: Can you see how the coefficients for `type` are related to each other across the two parametrizations of the model?

```
> LinearModel.4 <- lm(prestige ~ education + income + women + type,
+   data=Prestige)

> summary(LinearModel.4)

Call:
lm(formula = prestige ~ education + income + women + type, data = Prestige)

Residuals:
     Min       1Q   Median       3Q      Max
-14.7485  -4.4817   0.3119   5.2478  18.4978

Coefficients:
                    Estimate Std. Error t value Pr(>|t|)
(Intercept)        0.1821385  7.0466879   0.026 0.979435
education          3.6623557  0.6458300   5.671 1.63e-07 ***
income             0.0010428  0.0002623   3.976 0.000139 ***
women              0.0064434  0.0303781   0.212 0.832494
type[bc.v.others] -0.4980208  1.0194568  -0.489 0.626347
type[wc.v.prof]   -4.4111345  1.3968819  -3.158 0.002150 **
---
Signif. codes:  0 '***' 0.001 '**' 0.01 '*' 0.05 '.' 0.1 ' ' 1

Residual standard error: 7.132 on 92 degrees of freedom
Multiple R-squared:  0.8349,    Adjusted R-squared:  0.826
F-statistic: 93.07 on 5 and 92 DF,  p-value: < 2.2e-16
```

FIGURE 7.7: Output for the linear model prestige ~ income + education + women + type fit to the Prestige data, using customized contrasts for type.

7.3 Fitting Regression Splines and Polynomials*

The second formula toolbar in the *Linear Model* dialog makes it easy to add nonlinear *polynomial-regression* and *regression-spline* terms to a linear model.

7.3.1 Polynomial Terms

Some simple nonlinear relationships can be represented as low-order polynomials, such as a quadratic term, using regressors x and x^2 for a numeric explanatory variable x, or a cubic term, using x, x^2, and x^3. The resulting model is nonlinear in the explanatory variable x but linear in the parameters (the βs). R and the R Commander support both orthogonal and "raw" polynomials in linear model formulas.[18]

To add a polynomial term to the right-hand side of the model, single-click on a numeric variable in the *Variables* list box, and then press the appropriate toolbar button (either *orthogonal polynomial* or *raw polynomial*, as desired). There is a spinner in the *Linear Model* dialog for the degree of a polynomial term, and the default is 2 (i.e., a quadratic).

For example, inspection of the data (e.g., in a component-plus-residual plot, discussed in Section 7.8)[19] suggests that there may be a quadratic partial relationship between prestige and women in the regression of prestige on education, income, and women for the Canadian occupational prestige data.[20] I specify this quadratic relationship in the *Linear Model* dialog in Figure 7.8, using a raw second-degree polynomial, and producing the output in Figure 7.9. The quadratic coefficient in the model turns out not to be statistically significant ($p = 0.15$).

7.3.2 Regression Splines

Regression splines are flexible functions capable of representing a wide variety of nonlinear patterns in a model that, like a regression polynomial, is linear in the parameters. Both *B-splines* and *natural splines* are supported by the R Commander *Linear Model* dialog. Adding a spline term to the right-hand side of a linear model is similar to adding a polynomial term: The degrees of freedom for a spline term are controlled by a spinner (labelled *df*), with the default value 5; single-click on a variable in the list and then press the toolbar button for *B-spline* or *natural spline*.

For example, in Figure 7.10, I retain the quadratic specification for women, select education and press the *natural spline* button, and select income and press the *natural spline* button again. In both cases I leave the *df* spinner at its default value. These choices produce the model formula prestige $\sim$ poly(women, degree=2, raw=TRUE) + ns(education, df=5) + ns(income, df=5), regressing prestige on a quadratic in women and 5-*df* natural splines in education and income. The output for the resulting regression model isn't shown because the model requires graphical interpretation (see Section 7.6): The coefficient estimates for the regression splines are not simply interpretable.[21]

---

[18]The regressors of an orthogonal polynomial are uncorrelated, while those of a raw polynomial are just powers of the variable—for example, women and women^2. The fit of raw and orthogonal polynomials to the data is identical: They are just alternative parametrizations of the same regression. Raw polynomials may be preferred for simplicity of interpretation of the individual regression coefficients, but orthogonal polynomials tend to produce more numerically stable computations.

[19]Also see the discussion of partial residuals in effect plots in Section 7.6.

[20]To make this example a little simpler, I've omitted occupational type from the regression.

[21]In this revised model, however, where the partial relationships of prestige to education and income are modeled more adequately, the quadratic coefficient for women is statistically significant, with $p = 0.01$.

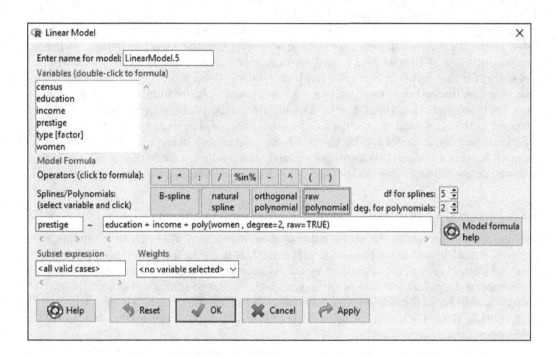

FIGURE 7.8: *Linear Model* dialog with a polynomial (quadratic) term for women in the regression of prestige on education, income, and women using the Prestige data set.

```
> LinearModel.5 <- lm(prestige ~ education + income + poly(women , degree=2,
+    raw=TRUE), data=Prestige)

> summary(LinearModel.5)

Call:
lm(formula = prestige ~ education + income + poly(women, degree = 2,
    raw = TRUE), data = Prestige)

Residuals:
     Min       1Q   Median       3Q      Max
 -20.2331  -5.3217   0.0987   4.9248  17.0059

Coefficients:
                                      Estimate Std. Error t value Pr(>|t|)
(Intercept)                          -6.1766164  3.2496931  -1.901   0.0603 .
education                             4.2601722  0.3899199  10.926  < 2e-16 ***
income                                0.0012720  0.0002778   4.579 1.39e-05 ***
poly(women, degree = 2, raw = TRUE)1 -0.1451676  0.0991716  -1.464   0.1465
poly(women, degree = 2, raw = TRUE)2  0.0015379  0.0010660   1.443   0.1523
---
Signif. codes:  0 '***' 0.001 '**' 0.01 '*' 0.05 '.' 0.1 ' ' 1

Residual standard error: 7.804 on 97 degrees of freedom
Multiple R-squared:  0.8024,    Adjusted R-squared:  0.7943
F-statistic: 98.48 on 4 and 97 DF,  p-value: < 2.2e-16
```

FIGURE 7.9: Output from the regression of prestige on education, income, and a quadratic in women for the `Prestige` data.

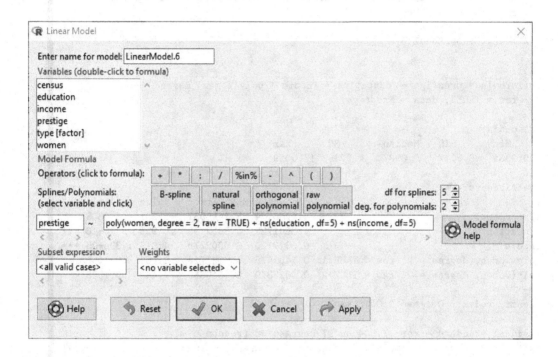

FIGURE 7.10: *Linear Model* dialog showing regression-spline and polynomial terms for the regression of `prestige` on `education`, `income`, and `women` in the `Prestige` data set.

7.4 Generalized Linear Models*

Briefly, *generalized linear models* (or *GLMs*), introduced in a seminal paper by Nelder and Wedderburn (1972), consist of three components:

1. A *random component* specifying the distribution of the response y conditional on explanatory variables. Traditionally, the random component is a member of an *exponential family*—the Gaussian (normal), binomial, Poisson, gamma, or inverse Gaussian families—but both the theory of generalized linear models and their implementation in R are now more general: In addition to the traditional exponential families, R provides for quasi-binomial and quasi-Poisson families that accommodate "over-dispersed" binomial and count data.

2. A *linear predictor*

$$\eta_i = \beta_0 + \beta_1 x_{1i} + \beta_2 x_{2i} + \cdots + \beta_k x_{ki}$$

    on which the expectation of the response variable $\mu_i = E(y_i)$ for the ith of n independent observations depends, where the regressors x_{ji} are prespecified functions of the explanatory variables—numeric explanatory variables, dummy regressors representing factors, interaction regressors, and so on, exactly as in the linear model.

3. A prespecified invertible *link function* $g(\cdot)$ that transforms the expectation of the response to the linear predictor, $g(\mu_i) = \eta_i$, and thus $\mu_i = g^{-1}(\eta_i)$. R implements identity, inverse, log, logit, probit, complementary log-log, square root, and inverse square links, with the applicable links varying by distributional family.

The most common GLM beyond the normal linear model (i.e., the Gaussian family paired with identity link) is the binomial logit model, suitable for dichotomous (two-category) response variables. For an illustration, I'll use data collected by Cowles and Davis (1987) on volunteering for a psychological experiment, where the subjects of the study were students in a university introductory psychology class.

The data for this example are contained in the data set `Cowles` in the **car** package,[22] which includes the following variables: `neuroticism`, a personality dimension with integer scores ranging from 0 to 24; `extraversion`, another personality dimension, also with scores from 0 to 24; `sex`, a factor with levels `"female"` and `"male"`; and `volunteer`, a factor with levels `"no"` and `"yes"`.

In analyzing the data, Cowles and Davis performed a logistic regression of volunteering on `sex` and the linear-by-linear interaction between `neuroticism` and `extraversion`. To fit Cowles and Davis's model, I first read the data from the **car** package in the usual manner, making `Cowles` the active data set in the R Commander. Then I select *Statistics > Fit models > Generalized linear model*, producing the dialog box in Figure 7.11.

The *Generalized Linear Model* dialog is very similar to the *Linear Model* dialog of the preceding section: The name of the model at the top (`GLM.7`) is automatically generated, and you can change it if you wish. Double-clicking on a variable in the list box enters the variable into the model formula. There are toolbars for entering operators, regression splines, and polynomials into the model formula, and there are boxes for subsetting the data set and for specifying prior weights.

---

[22]I'm grateful to Caroline Davis of York University for making the data available.

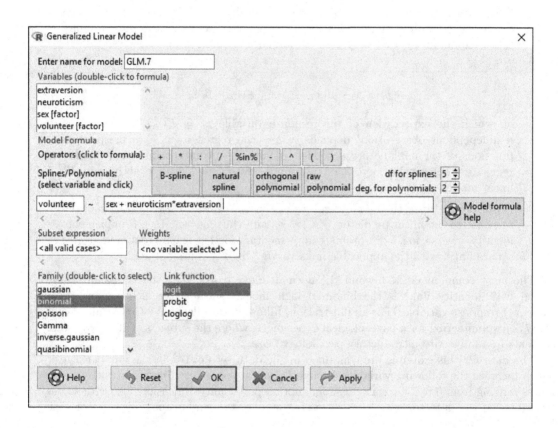

FIGURE 7.11: *Genealized Linear Model* dialog box for Cowles and Davis's logistic regression.

What's new in the *Generalized Linear Model* dialog are the *Family* and *Link function* list boxes, as are appropriate to a GLM. Families and links are coordinated: Double-clicking on a distributional family changes the available links. In each case, the *canonical link* for a particular family is selected by default. The initial selections are the *binomial* family and corresponding canonical *logit* link, which are coincidentally what I want for the example.

I proceed to complete the dialog by double-clicking on `volunteer` in the variable list, making it the response variable; then double-clicking on `sex` and on `neuroticism`; clicking the * button in the toolbar; and finally double-clicking on `neuroticism`—yielding the model formula `volunteer ~ sex + neuroticism*extraversion`. As in the *Linear Model* dialog, an alternative is to type the formula directly.

Appropriate responses for a binomial logit model include two-level factors (such as `volunteer` in the current example), logical variables (i.e., with values `FALSE` and `TRUE`), and numeric variables with two unique values (most commonly 0 and 1). In each case, the logit model is for the probability of the *second* of the two values—the probability that `volunteer` is `"yes"` in the example.

Clicking the *OK* button produces the output in Figure 7.12. The *Generalized Linear Model* dialog uses the R `glm` function to fit the model. The summary output for a generalized linear model is very similar to that for a linear model, including a table of estimated coefficients along with their standard errors, z values (*Wald statistics*) for testing that the coefficients are 0, and the two-sided p-values for these tests. For a logistic regression, the R Commander also prints the exponentiated coefficients, interpretable as multiplicative effects on the odds scale—here the odds of volunteering, $\Pr(\texttt{"yes"})/\Pr(\texttt{"no"})$.

The Wald z tests suggest a statistically significant interaction between `neuroticism` and `extraversion`, as Cowles and Davis expected, and a significant `sex` effect, with men less likely to volunteer than women who have equivalent scores on the personality dimensions. Because it's hard to grasp the nature of the interaction directly from the coefficient estimates, I'll return to this example in Section 7.6, where I'll plot the fitted model.

Although I've developed just one example of a generalized linear model in this section— a logit model for binary data—the R Commander *Generalized Linear Model* dialog is more flexible:

- The probit and complementary log-log (*cloglog*) link functions may also be used with binary data, as alternatives to the canonical logit link.

- The binomial family may also be used when the value of the response variable for each case (or *observation*) represents the proportion of "successes" in a given number of binomial trials, which may also vary by case. In this setting, the left-hand side of the model formula should give the proportion of successes, which could be computed as `successes/trials` (imagining that there are variables with these names in the active data set) directly in the left-hand box of the model formula, and the variable representing the number of trials for each observation (e.g., `trials`) should be given in the *Weights* box.

- Alternatively, for binomial data, the left-hand side of the model may be a two-column matrix specifying, respectively, the numbers of successes and failures for each observation, by typing, e.g., `cbind(successes, failures)` (again, imagining that these variable are in the active data set) into the left-hand-side box of the model formula.

- Other generalized linear models are specified by choosing a different family and corresponding link. For example, a Poisson regression model, commonly employed for count data, may be fit by selecting the *poisson* family and canonical *log* link (or, to get typically more realistic coefficient standard errors, by selecting the *quasipoisson* family with the *log* link).

```
> GLM.7 <- glm(volunteer ~ sex + neuroticism*extraversion,
+    family=binomial(logit), data=Cowles)

> summary(GLM.7)

Call:
glm(formula = volunteer ~ sex + neuroticism * extraversion,
    family = binomial(logit), data = Cowles)

Deviance Residuals:
    Min      1Q   Median      3Q      Max
-1.4749  -1.0602  -0.8934   1.2609   1.9978

Coefficients:
                          Estimate Std. Error z value Pr(>|z|)
(Intercept)              -2.358207   0.501320  -4.704 2.55e-06 ***
sex[T.male]              -0.247152   0.111631  -2.214  0.02683 *
neuroticism               0.110777   0.037648   2.942  0.00326 **
extraversion              0.166816   0.037719   4.423 9.75e-06 ***
neuroticism:extraversion -0.008552   0.002934  -2.915  0.00355 **
---
Signif. codes:  0 '***' 0.001 '**' 0.01 '*' 0.05 '.' 0.1 ' ' 1

(Dispersion parameter for binomial family taken to be 1)

    Null deviance: 1933.5  on 1420  degrees of freedom
Residual deviance: 1897.4  on 1416  degrees of freedom
AIC: 1907.4

Number of Fisher Scoring iterations: 4

> exp(coef(GLM.7))  # Exponentiated coefficients ("odds ratios")
            (Intercept)              sex[T.male]              neuroticism
             0.09458964               0.78102195               1.11714535
           extraversion neuroticism:extraversion
             1.18153740               0.99148400
```

FIGURE 7.12: Output from Cowles and Davis's logistic regression (volunteer ~ sex + neuroticism*extraversion).

7.5   Other Regression Models*

In addition to linear regression, linear models, and generalized linear models, the R Commander can fit *multinomial logit models* for categorical response variables with more than two categories (via *Statistics > Fit models > Multinomial logit model*), and *ordinal regression models* for ordered multi-category responses, including the *proportional-odds logit model* and the *ordered probit model* (*Statistics > Fit models > Ordinal regression model*). Although I won't illustrate these models here, many of the menu items in the *Models* menu apply to these classes of models. Moreover (as I will show in Chapter 9), R Commander plug-in packages can introduce additional classes of statistical models.

7.6   Visualizing Linear and Generalized Linear Models*

Introduced by Fox (1987), *effect plots* are graphs for visualizing complex regression models by focusing on particular explanatory variables or combinations of explanatory variables, holding other explanatory variables to typical values. One strategy is to focus successively on the explanatory variables in the *high-order terms* of the model—that is, terms that aren't marginal to others (see Section 7.2.2).

In the R Commander, effect displays can be drawn for linear, generalized linear, and some other statistical models via *Models > Graphs > Effect plots*. Figure 7.13 shows the resulting dialog box for Cowles and Davis's logistic regression from the previous section, GLM.7, which is the current statistical model in the R Commander. By default, the dialog offers to plot all high-order terms in the model—in this case, the sex main effect and the neuroticism-by-extraversion interaction. You may alternatively pick a subset of *Predictors* (explanatory variables) to plot.[23] For a linear or generalized linear model, there's also a check box for plotting partial residuals, unchecked by default, along with a slider for the span of a smoother fit to the residuals. Partial residuals and the accompanying smoother can be useful for judging departures from the functional form of the specified model, as I'll illustrate later in this section.

Clicking *OK* produces the graph in Figure 7.14: The left-hand panel shows the sex main effect, with neuroticism and extraversion set to average levels. The right-hand panel shows the neuroticism-by-extraversion interaction, for a group composed of males and females in proportion to their representation in the data set. In both graphs, the vertical volunteer axis is drawn on the logit scale but the tick-mark labels are on the estimated probability scale—that is, they represent the estimated probability of volunteering.[24]

In the plot of the interaction, the horizontal axis of each panel is for neuroticism, while extraversion takes on successively larger values across its range, from the lower-left panel to the upper-right panel. The value of extraversion for each panel is represented by the small vertical line in the strip labelled *extraversion* at the top of the panel.

---

[23]You're not constrained to select focal explanatory variables that correspond to high-order terms in the model, or even to terms in the model. For example, for Cowles and Davis's logistic regression, you could select all three explanatory variables, sex, neuroticism, and extraversion, even though the three-way interaction among these variables isn't in the model. In this case, the effect plot would be drawn for combinations of the values of the three explanatory variables.

[24]This strategy is used in general for effect plots of generalized linear models in the R Commander: The vertical axis is drawn on the scale of the linear predictor—the scale on which the model is linear—but labelled on the generally more interpretable scale of the response.

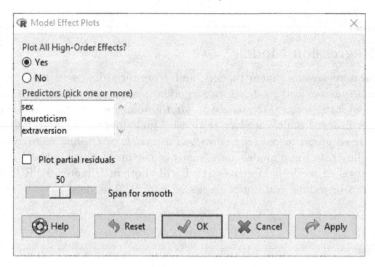

FIGURE 7.13: *Model Effect Plots* dialog box for Cowles and Davis's logistic regression
(`volunteer ~ sex + neuroticism*extraversion`).

The lines in the panels represent the combined effect of `neuroticism` and `extraversion`,
and are computed using the estimated coefficients for the `neuroticism:extraversion` interaction along with the coefficients for the `neuroticism` and `extraversion` "main effects,"
which are marginal to the interaction. It's clear that there's a positive relationship between
volunteering and `neuroticism` at the lowest level of `extraversion`, but that this relationship becomes negative at the highest level of `extraversion`.

The error bars in the effect plot for `sex` and the gray bands in the plot of the `neuroti-
cism-by-extraversion` interaction represent point-wise 95% confidence intervals around
the estimated effects. The "rug-plot" at the bottom of each panel in the display of the `neu-
roticism-by-extraversion` interaction shows the marginal distribution of `neuroticism`,
with the lines slightly displaced to decrease over-plotting. The rug-plot isn't terribly useful
here, because `neuroticism` just takes on integer scores.

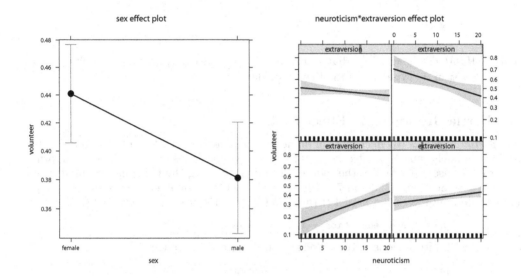

FIGURE 7.14: Effect plots for the high-order terms in Cowles and Davis's logistic regression (volunteer ~ sex + neuroticism*extraversion). The graphs are shown in monochrome; they (and the other effect plots in the chapter) were originally in color.

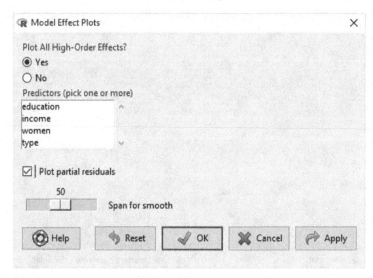

FIGURE 7.15: *Model Effect Plots* dialog box for `LinearModel.2` (`prestige ~ education + income + women + type`) fit to the `Prestige` data.

7.6.1   Partial Residuals in Effect Plots

Adding *partial residuals* to effect plots of numeric explanatory variables in linear and generalized linear models can be an effective tool for judging departures from the functional form (linear or otherwise) specified in the model. I'll illustrate using the Canadian occupational prestige data. In Sections 7.2.3 and 7.3, I fit several models to the `Prestige` data, including an additive dummy-regression model (`LinearModel.2` in Figure 7.4 on page 136),

           prestige ~ education + income + women + type

and a model with interactions (`LinearModel.3` in Figure 7.5 on page 137),

           prestige ~ type*education + type*income

among others.

The R Commander session in this chapter is unusual in that I've read three data sets (`Duncan`, `Prestige`, and `Cowles`) and fit statistical models to each. It is much more common to work with a single data set in a session. Nevertheless, as I explained, the R Commander allows you to switch among models and data sets, and takes care of synchronizing models with the data sets to which they were fit. After making `LinearModel.2` the active model, I return to the *Model Effect Plots* dialog, displayed in Figure 7.15. I check *Plot partial residuals* and click *OK*, producing Figure 7.16. Partial residuals are plotted for the numeric predictors but not for the factor `type`; this is reflected in a warning printed in the *Messages* pane, which I'll simply ignore.

The solid lines in the effect plots represent the model fit to the data, while the broken lines are smooths of the partial residuals. If the lines for the fitted model and smoothed partial residuals are similar, that lends support to the specified functional form of the model. The partial residuals are computed by adding the residual for each observation to the line representing the fitted effect. It appears as if the `education` effect is modelled reasonably, but the `income` and `women` effects appear to be nonlinear.

`LinearModel.3` includes interactions between `type` and each of `education` and `income`. Figure 7.17 shows effect plots with partial residuals for the high-order terms in this model.

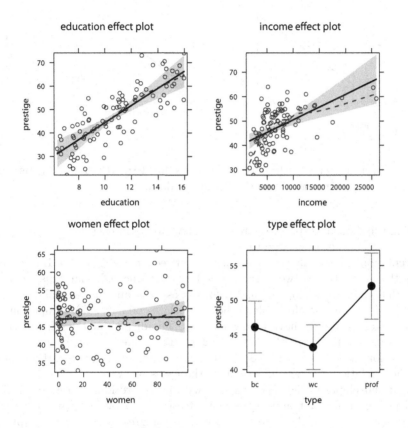

FIGURE 7.16: Effect displays with partial residuals for LinearModel.2 (prestige ~ education + income + women + type) fit to the Prestige data.

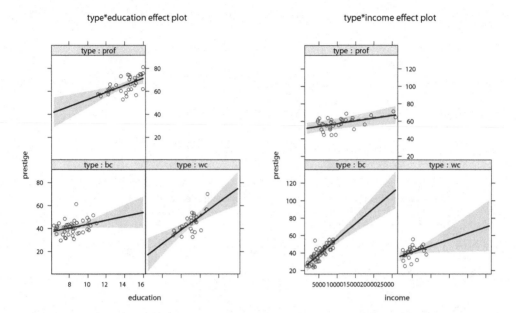

FIGURE 7.17: Effect displays with partial residuals for `LinearModel.3` (prestige ~ `type*education + type*income`) fit to the `Prestige` data.

Because dividing the data by `type` leaves relatively few points in each panel of the plots, I set the span of the smoother to a large value, 0.9.[25]

The apparent nonlinearity in the relationship between `prestige` and `income` is accounted for by the interaction between `income` and `type`:[26] The right-hand display of Figure 7.17 shows that the `income` slope is smaller for professional and managerial occupations (i.e., `type = "prof"`) than for blue-collar (`"bc"`) or white-collar (`"wc"`) occupations, and professional occupations tend to have higher incomes. The display at the left, for the education-by-`type` interaction, suggests that the `education` slope is steeper for white-collar occupations than for the other types of occupations. The smooths of the partial residuals indicate that these relationships are linear within the levels of `type`.

The confidence envelopes in the effect displays with partial residuals in Figure 7.17 also make a useful pedagogical point about precision of estimation of the regression surface: Where data are sparse—or, in the extreme, absent—the regression surface is imprecisely estimated.

`LinearModel.6`, fit in Section 7.3,

`prestige ~ poly(women, degree=2, raw=TRUE) + ns(education, df=5) + ns(income, df=5)`

uses a quadratic in `women` along with regression splines for `income` and `education`, which should capture the unmodelled nonlinearity observed in Figure 7.16; the model doesn't include the factor `type`, however. I make `LinearModel.6` the active model and repeat the effect plots, which are shown in Figure 7.18. Here, the fitted model and smoothed residuals agree well with each other.

---

[25]Smoothing scatterplots is discussed in Section 5.4.

[26]Recall, however, that the explanatory variable `women` isn't included in this model.

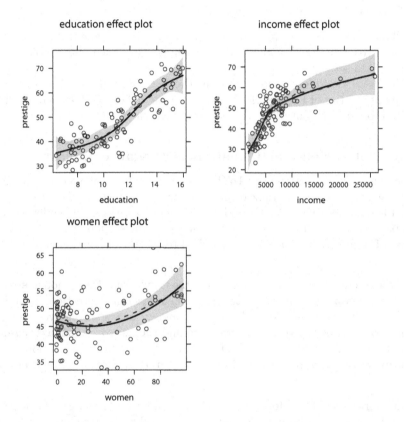

FIGURE 7.18: Effect displays with partial residuals for LinearModel.6 (prestige ∼ ns(education, df=5) + ns(income, df=5) + poly(women, degree=2, raw=TRUE)) fit to the Prestige data.

7.7   Confidence Intervals and Hypothesis Tests for Coefficients

The *Models* menu includes several menu items for constructing confidence intervals and performing hypothesis tests for regression coefficients. As I explained, tests for individual coefficients in linear and generalized linear models appear in the model summaries. This section describes how to perform more elaborate tests, for example, for a related subset of coefficients.

7.7.1   Confidence Intervals

To illustrate, I'll make Duncan's occupational prestige regression (`RegModel.1` in Figure 7.2 on page 131) the active statistical model in the R Commander, which automatically makes `Duncan` the active data set.

Selecting *Models > Confidence intervals* from the R Commander menus leads to the simple dialog box at the top of Figure 7.19.[27] Retaining the default 0.95 level of confidence and clicking *OK* produces the output at the bottom of the figure.

7.7.2   Analysis of Variance and Analysis of Deviance Tables*

You can compute an *analysis of variance (ANOVA) table* for a linear model or an analogous *analysis of deviance table* for a generalized linear model via *Models > Hypothesis tests > ANOVA table*. I'll illustrate with Cowles and Davis's logistic regression model (`GLM.7` in Figure 7.12 on page 148), selecting it as the active model in the session. The *ANOVA Table* dialog at the top of Figure 7.20 offers three "types" of tests, conventionally named *Types I, II*, and *III*:

- In addition to the intercept, there are four terms in the Cowles and Davis model: `sex`, `neuroticism`, `extraversion`, and the `neuroticism:extraversion` interaction. Type I tests are *sequential*, and thus test (in a short-hand terminology) `sex` ignoring everything else; `neuroticism` after `sex` but ignoring `extraversion` and the `neuroticism:extraversion` interaction; `extraversion` after `sex` and `neuroticism` ignoring the interaction; and the `neuroticism:extraversion` interaction after all other terms. Sequential tests are rarely sensible.

- Type II tests are formulated in conformity with the principle of marginality (Section 7.2.2): `sex` after all other terms, including the `neuroticism:extraversion` interaction; `neuroticism` after `sex` and `extraversion` but ignoring the `neuroticism:extraversion` interaction to which the `neuroticism` "main effect" is marginal; similarly, `extraversion` after `sex` and `neuroticism` but ignoring the `neuroticism:extraversion` interaction; and `neuroticism:extraversion` after all the other terms. More generally, each term is tested ignoring terms to which it is marginal (i.e., ignoring its higher-order relatives). This is generally a sensible approach and is the default in the dialog.

Type II tests for Cowles and Davis's logistic regression are shown at the bottom of Figure 7.20. For a generalized linear model like Cowles and Davis's logistic regression, the R Commander computes an analysis of deviance table with likelihood ratio tests, which entails refitting the model (in some instances twice) for each test. In this example, each

---

[27]For a generalized linear model, the *Confidence Intervals* dialog provides an option to base confidence intervals on either the likelihood ratio statistic or on the Wald statistic. The former requires more computation, but it is the default because confidence intervals based on the likelihood ratio statistic tend to be more accurate.

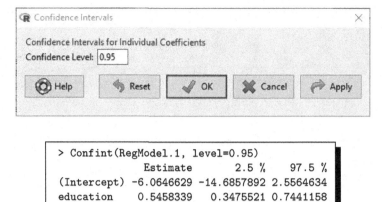

FIGURE 7.19: *Confidence Intervals* dialog and resulting output for Duncan's occupational prestige regression (`prestige ~ education + income`).

likelihood ratio chi-square test has one degree of freedom because each term in the model is represented by a single coefficient. Because the `neuroticism:extraversion` interaction is highly statistically significant, I won't interpret the tests for the `neuroticism` and `extraversion` main effects, which assume that the interaction is nil. The `sex` effect is also statistically significant.

- Type III tests are for each term after all of the others. This isn't sensible for Cowles and Davis's model, although it might be rendered sensible, for example, by centering each of `neuroticism` and `extraversion` at their means prior to fitting the model. Even where Type III tests can correspond to sensible hypotheses, such as in analysis of variance models, they require careful formulation in R.[28] I suggest that you avoid Type III tests unless you know what you're doing.

---

[28]For Type III tests to address sensible hypotheses, the contrasts used for factors in an ANOVA model must be orthogonal in the basis of the design. The functions `contr.Sum`, `contr.Helmert`, and `contr.poly` produce contrasts with this property, but the R Commander default dummy-coded `contr.Treatment` does not. For this reason, the R Commander *Multi-Way Analysis of Variance* dialog (Section 6.1.3) uses `contr.Sum` to fit ANOVA models (see, in particular, Figures 6.9 and 6.10, pages 118–119), and so the resulting models can legitimately employ Type III tests.

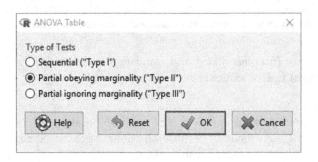

```
> Anova(GLM.7, type="II", test="LR")
Analysis of Deviance Table (Type II tests)

Response: volunteer
                       LR Chisq Df Pr(>Chisq)
sex                      4.9184  1   0.026572 *
neuroticism             0.3139  1   0.575316
extraversion           22.1372  1  2.538e-06 ***
neuroticism:extraversion 8.6213  1   0.003323 **
---
Signif. codes:  0 '***' 0.001 '**' 0.01 '*' 0.05 '.' 0.1 ' ' 1
```

FIGURE 7.20: *ANOVA Table* dialog and resulting Type II tests for Cowles and Davis's logistic regression (volunteer ~ sex + neuroticism*extraversion).

```
> LinearModel.8 <- lm(prestige ~ I(education + income), data=Duncan)

> summary(LinearModel.8)

Call:
lm(formula = prestige ~ I(education + income), data = Duncan)

Residuals:
    Min     1Q  Median     3Q     Max
-29.605  -6.708  0.124   6.979  33.289

Coefficients:
                      Estimate Std. Error t value Pr(>|t|)
(Intercept)           -6.06319    4.22540  -1.435    0.159
I(education + income)  0.56927    0.03958  14.382   <2e-16 ***
---
Signif. codes:  0 '***' 0.001 '**' 0.01 '*' 0.05 '.' 0.1 ' ' 1

Residual standard error: 13.22 on 43 degrees of freedom
Multiple R-squared:  0.8279,    Adjusted R-squared:  0.8239
F-statistic: 206.8 on 1 and 43 DF,  p-value: < 2.2e-16
```

FIGURE 7.21: Summary of a regression model fit to Duncan's occupational prestige data forcing the education and income coefficients to be equal (prestige ~ I(education + income)).

7.7.3 Tests Comparing Models*

The R Commander allows you to compute a likelihood ratio F test or chi-square test for two regression models, one of which is nested within the other.[29] To illustrate, I return to the Duncan data set and fit a version of Duncan's regression that sets the coefficients of education and income equal to each other, specifying the linear-model formula prestige ~ I(income + education).[30] This at least arguably makes sense in that both explanatory variables are percentages—respectively of high school graduates and of high-income earners. The resulting model, LinearModel.8, is summarized in Figure 7.21.

Picking *Models > Hypothesis tests > Compare two models* leads to the dialog at the top of Figure 7.22. I select the more general RegModel.1 as the first model and the more specific, constrained, LinearModel.8 as the second model, but the order of the selections is immaterial—the same F test is produced in both cases. Clicking *OK* in the dialog box results in the output at the bottom of Figure 7.22. The hypothesis of equal coefficients for education and income is plausible, $p = 0.79$—after all, the two coefficients are quite similar in the original regression (Figure 7.2 on page 131), $b_{education} = 0.55$ and $b_{income} = 0.60$.

---

[29] F tests are computed for linear models and for generalized linear models (such as quasi-Poisson models) for which there's an estimated dispersion parameter; chi-square tests are computed for generalized linear models (such as binomial models) for which the dispersion parameter is fixed. The same is true of analysis of variance and analysis of deviance tables.

[30] Recall that the identity (or inhibit) function I is needed here so that + is interpreted as addition.

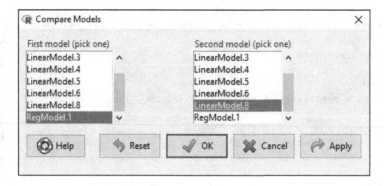

```
> anova(RegModel.1, LinearModel.8)
Analysis of Variance Table

Model 1: prestige ~ education + income
Model 2: prestige ~ I(education + income)
  Res.Df    RSS Df Sum of Sq      F Pr(>F)
1     42 7506.7
2     43 7518.9 -1   -12.195 0.0682 0.7952
```

FIGURE 7.22: *Compare Models* dialog and resulting output for Duncan's occupational prestige regression, testing the equality of the education and income regression coefficients.

7.7.4 Testing Linear Hypotheses*

The menu selection *Models > Hypothesis tests > Linear hypothesis* allows you to formulate and test general linear hypotheses about the coefficients in a regression model. To illustrate, I'll again use Duncan's occupational prestige regression, RegModel.1. Figures 7.23 and 7.24 show the *Test Linear Hypothesis* dialog set up to test two different hypotheses, along with the corresponding output:

1. In Figure 7.23, H_0: $1 \times \beta_{\text{education}} - 1 \times \beta_{\text{income}} = 0$ (i.e., H_0: $\beta_{\text{education}} = \beta_{\text{income}}$). This is the same hypothesis that I tested by the model-comparison approach immediately above, and of course it produces the same F test.

2. In Figure 7.24, H_0: $1 \times \beta_{\text{education}} = 0$, $1 \times \beta_{\text{income}} = 0$, which is equivalent to, and thus produces the same F test as, the omnibus null hypothesis in the linear-model summary output, H_0: $\beta_{\text{education}} = \beta_{\text{income}} = 0$ (see Figure 7.2 on page 131). Because the linear hypothesis consists of two equations, the F statistic for the hypothesis has two *df* in the numerator.

There may be up to as many equations in a linear hypothesis as the number of coefficients in the model, with the number of equations controlled by the slider at the top of the dialog. The equations must be linearly independent of one another—that is, they may not be redundant. Initially, all of the cells in each row are 0, including the cell representing the right-hand side of the hypothesis, which is usually left at 0. The *Test Linear Hypothesis* dialog for a linear model provides for an optional "sandwich" estimator of the coefficient covariance matrix, which may be used to adjust statistical inference for autocorrelated or heteroscedastic errors (nonconstant error variance).

```
> local({
+     .Hypothesis <- matrix(c(0,1,-1), 1, 3, byrow=TRUE)
+     .RHS <- c(0)
+     linearHypothesis(RegModel.1, .Hypothesis, rhs=.RHS)
+ })
Linear hypothesis test

Hypothesis:
education - income = 0

Model 1: restricted model
Model 2: prestige ~ education + income

  Res.Df    RSS Df Sum of Sq      F Pr(>F)
1     43 7518.9
2     42 7506.7  1    12.195 0.0682 0.7952
```

FIGURE 7.23: Testing a linear hypothesis for Duncan's occupational prestige regression: H_0: $\beta_{\text{education}} = \beta_{\text{income}}$.

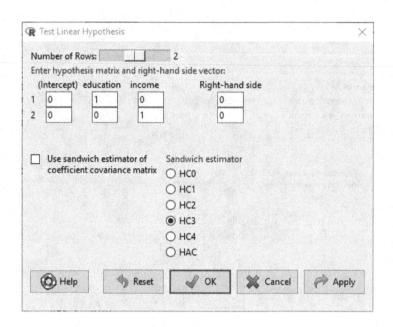

```
> local({
+    .Hypothesis <- matrix(c(0,1,0,0,0,1), 2, 3, byrow=TRUE)
+    .RHS <- c(0,0)
+    linearHypothesis(RegModel.1, .Hypothesis, rhs=.RHS)
+ })
Linear hypothesis test

Hypothesis:
education = 0
income = 0

Model 1: restricted model
Model 2: prestige ~ education + income

  Res.Df    RSS Df Sum of Sq      F    Pr(>F)
1     44  43688
2     42   7507  2     36181 101.22 < 2.2e-16 ***
---
Signif. codes:  0 '***' 0.001 '**' 0.01 '*' 0.05 '.' 0.1 ' ' 1
```

FIGURE 7.24: Testing a linear hypothesis for Duncan's occupational prestige regression: H_0: $\beta_{\text{education}} = \beta_{\text{income}} = 0$.

7.8   Regression Model Diagnostics*

Regression diagnostics are methods for determining whether a regression model that's been fit to data adequately summarizes the data. For example, is a relationship that's assumed to be linear actually linear? Do one or a small number of influential cases unduly affect the results?

Many standard methods of regression diagnostics are implemented in the R Commander *Models > Numerical diagnostics* and *Models > Graphs* menus—indeed, too many to cover in detail in this already long chapter. Luckily, most of the diagnostics dialogs are entirely straightforward, and some of the diagnostics menu items produce results directly, without invoking a dialog box. I'll illustrate with Duncan's occupational prestige regression (RegModel.1 in Figure 7.2 on page 131). As usual, I assume that the statistical methods covered here are familiar. Regression diagnostics are taken up in many regression texts; see, in particular, Fox (2016), Weisberg (2014), Cook and Weisberg (1982), or (for a briefer treatment) Fox (1991).

The numerical diagnostics available in the R Commander include *generalized variance-inflation factors* (Fox and Monette, 1992) for diagnosing collinearity in linear and generalized linear models, the *Breusch–Pagan test* for nonconstant error variance in a linear model (Breusch and Pagan, 1979), independently proposed by Cook and Weisberg (1983), the *Durbin–Watson test* for autocorrelated errors in linear time-series regression (Durbin and Watson, 1950, 1951), the *RESET test* for nonlinearity in a linear model (Ramsey, 1969), and a *Bonferonni outlier test* based on the *studentized residuals* from a linear or generalized linear model (see, e.g., Fox, 2016, Chapter 11).

I'll illustrate with *Models > Numerical diagnostics > Breusch-Pagan test for heteroscedasticity*, leading to the dialog at the top of Figure 7.25. The default is to test for error variance that increases (or decreases) with the level of the response (through the *Fitted values*), but the dialog is flexible and accommodates dependence of the error variance on the explanatory variables or on a linear predictor based on any variables in the data set. I leave the dialog at its default, producing the output at the bottom of Figure 7.25. There is, therefore, no evidence that the variance of the errors in Duncan's regression depends on the level of the response.

There are also many graphical diagnostics available through the R Commander: Basic diagnostic plots produced by the R `plot` function applied to a linear or generalized linear model; *residual quantile-comparison plots*, for example to diagnose non-normal errors in a linear model; *component-plus-residual (partial-residual) plots* for nonlinearity in additive linear or generalized linear models[31]; *added-variable plots* for diagnosing unusual and influential data in linear and generalized linear models; and an *"influence plot"*—a diagnostic graph that simultaneously displays studentized residuals, *hat-values (leverages)*, and *Cook's distances*.

I'll selectively demonstrate these graphical diagnostics by applying an influence plot, added-variable plots, and component-plus-residual plots to Duncan's regression (RegModel.1), making it the active model. I invite the reader to explore the other diagnostics as well.

Selecting *Models > Graphs > Influence plot* produces the dialog box at the top of Figure 7.26. I've left all of the selections in the dialog at their default values, including automatic identification of unusual points:[32] Two cases are selected from each of the most extreme stu-

---

[31]In Section 7.6, I showed how to add partial residuals to effect displays. For an additive model, that approach produces traditional component-plus-residual plots, but it is more flexible in that it can be applied as well to models with interactions.

[32]The default *Automatic* point identification has the advantage of working in the R Markdown document produced by the R Commander. As explained in Section 5.4.4, graphs that require direct interaction are not included in the R Markdown document.

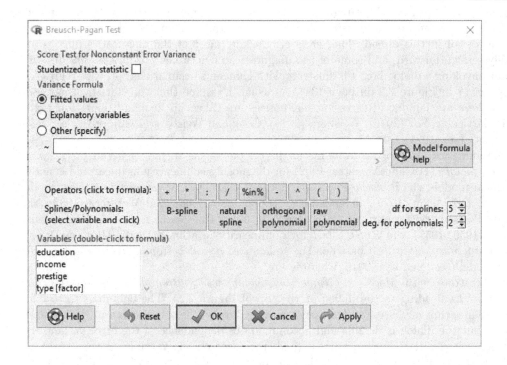

FIGURE 7.25: *Breusch-Pagan Test* dialog and resulting output for Duncan's occupational prestige regression (prestige ~ education + income).

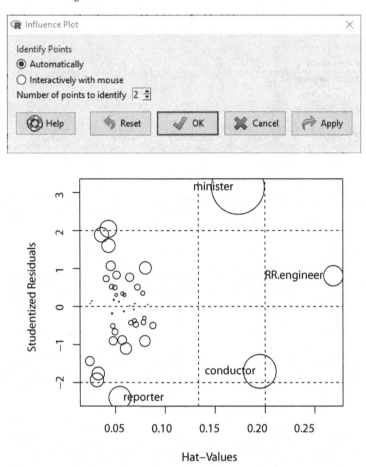

FIGURE 7.26: *Influence Plot* dialog and resulting graph for Duncan's occupational prestige regression (`prestige ~ education + income`).

dentized residuals, hat-values, and Cook's distances, potentially identifying up to six points (although this is unlikely to happen because influential points combine high leverage with a large residual). The resulting graph appears at the bottom of the figure, and, as it turns out, four relatively unusual points are identified: The occupation `RR.engineer` (railroad engineer) is at a high leverage point but has a small studentized residual; `reporter` has a relatively large (negative) studentized residual but small leverage; (railroad) `conductor` and, particularly, `minister` have comparatively large studentized residuals and moderately high leverage. The areas of the circles are proportional to Cook's influence measure, so `minister`, combining a large residual with fairly high leverage, has the most influence on the regression coefficients.

Models > Graphs > Added-variable plots leads to the dialog box at the top of Figure 7.27; as before, the dialog allows the user to select a method for identifying noteworthy points in the resulting graphs. Once again, I retain the default automatic point identification but now

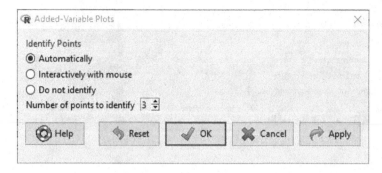

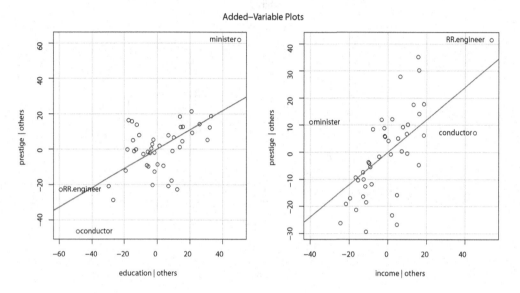

FIGURE 7.27: *Added-Variable Plots* dialog and resulting graphs for Duncan's occupational prestige regression (`prestige ~ education + income`).

increase the number of points to be identified in each graph from the default two to three.[33] Clicking *OK* produces the graphs at the bottom of Figure 7.27. The slope of the least-squares line in each added-variable plot is the coefficient for the corresponding explanatory variable in the multiple regression, and the plot shows how the cases influence the coefficient—in effect, transforming the multiple regression into a series of simple regressions, each with other explanatory variables controlled.

The added-variable plots are even more informative than the influence plot: The occupations `minister` and `conductor` appear to be an *influential pair*, inflating the `education` coefficient and decreasing the `income` coefficient. The occupation `RR.engineer` has high leverage on both coefficients but is more or less in line with the rest of the data.[34]

Selecting *Models > Graphs > Component+residual plots* brings up the dialog box at the top of Figure 7.28. Because there are only 45 cases in the `Duncan` data set, I increase the

---

[33]If you want to experiment with automatic identification of differing numbers of points, press the *Apply* button rather than *OK*.

[34]I'll leave it as an exercise for the reader to remove the occupations `minister` (case 6) and `conductor` (case 16) from the data and refit the regression—most conveniently via the *Subset expression* box in the *Linear Model* or *Linear Regression* dialog; you can use the subset -c(6, 16).

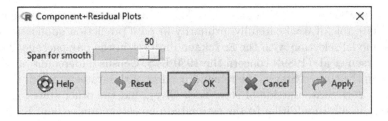

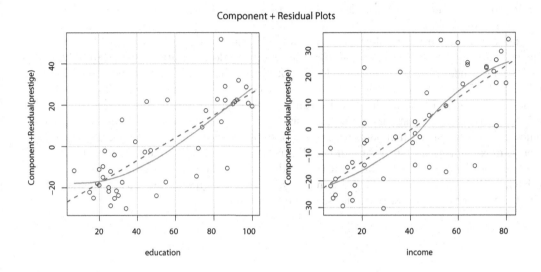

FIGURE 7.28: *Component+Residual Plots* dialog and resulting graphs for Duncan's occupational prestige regression (`prestige ~ education + income`).

span for the smoother from the default 50 percent to 90 percent. The resulting component-plus-residual plots, at the bottom of Figure 7.28, suggest that the partial relationships of `prestige` to `education` and `income` in Duncan's data are reasonably linear.

Finally, in the *Models* menu, *Add observation statistics to data* allows you to add fitted values (i.e., $\hat{y}$), residuals, studentized residuals, hat-values, Cook's distances, and observations indices $(1, 2, \ldots, n)$ to the active data set. These quantities, which (with the exception of observation indices) are named for the model to which they belong (e.g., `residuals.RegModel.1`), may then be used, for example, to create customized diagnostic graphs, such as an index plot of Cook's distances versus observation indices.

7.9 Model Selection*

The R Commander *Models* menu includes modest facilities for comparing regression models and for automatic model selection. The menu items *Models > Akaike Information Criterion (AIC)* and *Models > Bayesian Information Criterion (BIC)* print the AIC or BIC model selection statistics for the current statistical model. *Models > Stepwise model selection* performs stepwise regression for a linear or generalized linear model, while *Models > Subset model selection* performs all-subsets regression for a linear model. Although I've never been terribly enthusiastic about automatic model selection methods, I believe that these methods do have a legitimate role, if used carefully, primarily in pure prediction applications.

I'll illustrate model selection with the `Ericksen` data set in the **car** package. The data, described by Ericksen et al. (1989), concern the 1980 U. S. Census undercount, and pertain to 16 large cities in the United States, the remaining parts of the states to which these cities belong (e.g., New York State outside of New York City), and the other states. There are, therefore, 66 cases in all. In addition to the estimated percentage undercount in each area, the data set contains a variety of characteristics of the areas. A linear least-squares regression of `undercount` on the other variables in the data set reveals that some of the predictors are substantially collinear.[35] I fit this initial linear model with the formula `undercount ~ .` and obtained variance-inflation factors for the regression coefficients via *Models > Numerical diagnostics > Variance-inflation factors*. The relevant output is in Figure 7.29.

Choosing *Models > Subset model selection* produces the dialog box at the top of Figure 7.30. All of the options in this dialog remain at their defaults, including using the BIC for model selection. Pressing the *OK* button produces the graph at the bottom of Figure 7.30, plotting the "best" model of each size $k = 1, \ldots, 9$, according to the BIC. The predictors included in each model are represented by filled-in squares, and smaller (i.e., larger-in-magnitude negative) values of the BIC represent "better" models. Notice that the regression intercept is included in all of the models. The best model overall, according to the BIC, includes the four predictors `minority`, `crime`, `language`, and `conventional`. I invite the reader to try stepwise model selection as an alternative.[36]

---

[35]Ericksen et al. (1989) performed a more sophisticated weighted-least-squares regression.

[36]All-subsets regression in the R Commander is performed by the `regsubsets` function in the **leaps** package (Lumley and Miller, 2009), while stepwise regression is performed by the `stepAIC` function in the **MASS** package (Venables and Ripley, 2002). Although the distinction is not relevant to this example, where the full model is additive and all terms have one degree of freedom, `stepAIC` respects the structure of the model—for example, keeping dummy variables for a factor together and considering only models that obey the principle of marginality—while `regsubsets` does not.

```
> LinearModel.9 <- lm(undercount ~ ., data=Ericksen)

> summary(LinearModel.9)

Call:
lm(formula = undercount ~ ., data = Ericksen)

Residuals:
    Min      1Q  Median      3Q     Max
-2.8356 -0.8033 -0.0553  0.7050  4.2467

Coefficients:
                Estimate Std. Error t value Pr(>|t|)
(Intercept)    -0.611411   1.720843  -0.355 0.723678
minority        0.079834   0.022609   3.531 0.000827 ***
crime           0.030117   0.012998   2.317 0.024115 *
poverty        -0.178369   0.084916  -2.101 0.040117 *
language        0.215125   0.092209   2.333 0.023200 *
highschool      0.061290   0.044775   1.369 0.176415
housing        -0.034957   0.024630  -1.419 0.161259
city[T.state]  -1.159982   0.770644  -1.505 0.137791
conventional    0.036989   0.009253   3.997 0.000186 ***
---
Signif. codes:  0 '***' 0.001 '**' 0.01 '*' 0.05 '.' 0.1 ' ' 1

Residual standard error: 1.426 on 57 degrees of freedom
Multiple R-squared:  0.7077,    Adjusted R-squared:  0.6667
F-statistic: 17.25 on 8 and 57 DF,  p-value: 1.044e-12

> vif(LinearModel.8)
    minority       crime      poverty     language   highschool      housing
    5.009065    3.343586     4.625178     1.635568     4.619169     1.871745
        city conventional
    3.537750     1.691320
```

FIGURE 7.29: Regression output and variance-inflation factors for Ericksen et al.'s Census undercount data, fitting the linear model undercount ~ ..

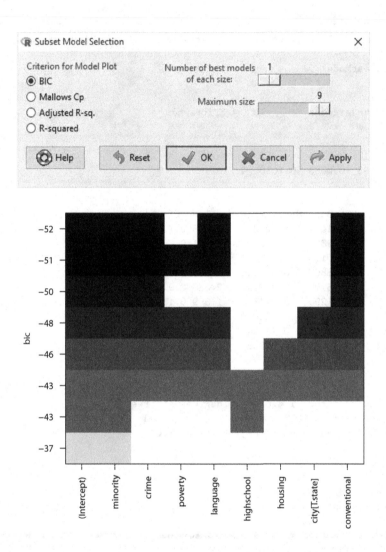

FIGURE 7.30: *Subset Model Selection* dialog and resulting graph for Ericksen et al.'s Census undercount data.

8

Probability Distributions and Simulation

This chapter explains how to use the R Commander to perform computations on probability distributions, to graph probability distributions, and to conduct simple random simulations.

8.1 Working with Probability Distributions

You can throw away your statistical tables because the R Commander has extensive facilities for computing probabilities and quantiles for 13 continuous and 5 discrete distributions used in statistics, well beyond what's typically required for a basic statistics course. The sub-menus and menu items for probability computations are under the R Commander *Distributions* menu (see Figure A.8 on page 205) and include the distributions shown in Table 8.1. In addition to these probability distributions, the *Distributions* menu includes an item for setting R's random-number generator seed (discussed in Section 8.3). None of the menu items in the *Distributions* menu requires an active data set, but the menu items for sampling from distributions *create* simulated data sets.

8.1.1 Continuous Distributions

Each continuous distribution in the *Distributions* menu has menu items for computing cumulative and tail probabilities, computing quantiles, generating random samples (discussed in Section 8.3), and plotting density and distribution functions (discussed in Section 8.2). The dialogs that implement this functionality have a common general format, and so I'll illustrate with a few representative cases.

The *Normal Probabilities* dialog, produced by selecting *Distributions > Continuous distributions > Normal distribution > Normal probabilities*, appears in Figure 8.1, with the following default selections, shown at the top of the figure: *Variable value(s)* are initially unspecified; the *Mean* is set to 0 and the *Standard deviation* to 1; and the *Lower tail* radio button is selected.[1] Entries in the *Variable values(s)* box must be separated by commas, spaces, or both. At the bottom of the figure, I enter the *Variable value(s)* 55 70 85 100 115 130 145, set the *Mean* to 100 and the *Standard deviation* to 15 (as is typical of IQ scores), press the *Upper tail* radio button, and click *OK*, obtaining

```
> pnorm(c(55,70,85,100,115,130,145), mean=100, sd=15, lower.tail=FALSE)
[1] 0.998650102 0.977249868 0.841344746 0.500000000 0.158655254 0.022750132
[7] 0.001349898
```

Thus, for example, for $X \sim N(100, 15^2)$, $\Pr(X \geq 145) = 0.001349898$, and so approximately 0.1% of values are as large as $X = 145$ or larger.

Choosing *Distributions > Continuous distributions > Normal distribution > Normal quantiles* from the R Commander menus brings up the simple dialog box in Figure 8.2.

---

[1]With *Lower tail* selected, the dialog returns values from the *cumulative distribution function (CDF)*.

TABLE 8.1: Continuous and discrete distributional families in the R Commander *Distributions* menu.

Continuous Distributions	Discrete Distributions
Normal (Gaussian)	Binomial
t	Poisson
Chi-square	Geometric
F	Hypergeometric
Exponential	Negative binomial
Uniform	
Beta	
Cauchy	
Logistic	
Log-normal	
Gamma	
Weibull	
Gumbel	

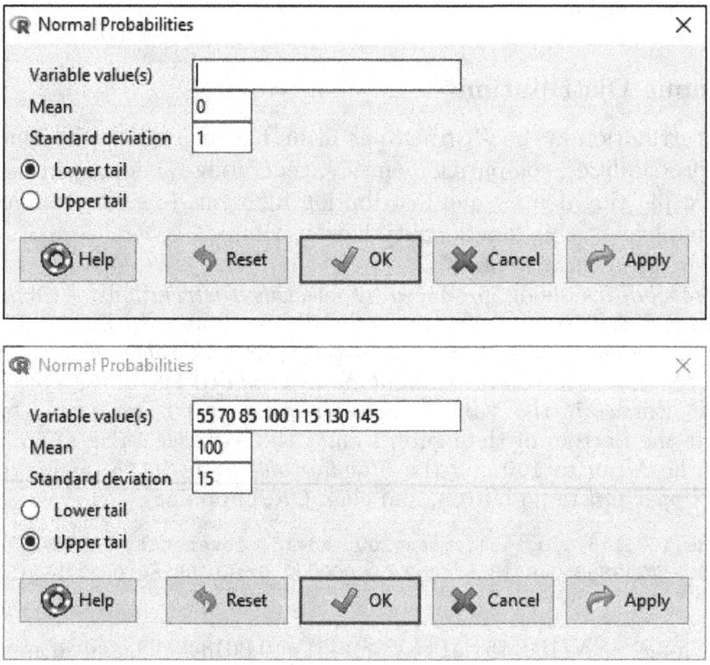

FIGURE 8.1: The *Normal Probabilities* dialog box: initial state (top); completed (bottom).

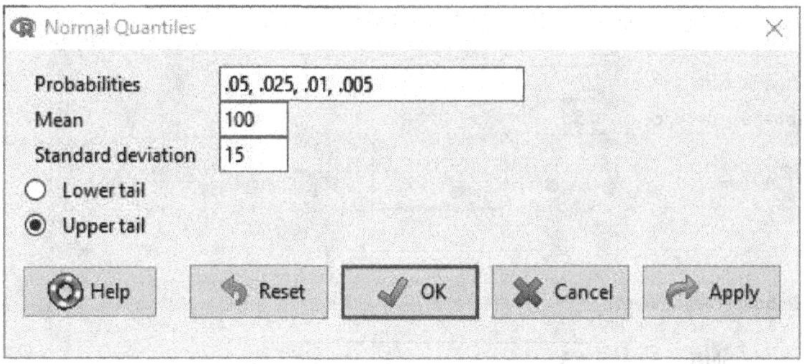

FIGURE 8.2: The *Normal Quantiles* dialog box.

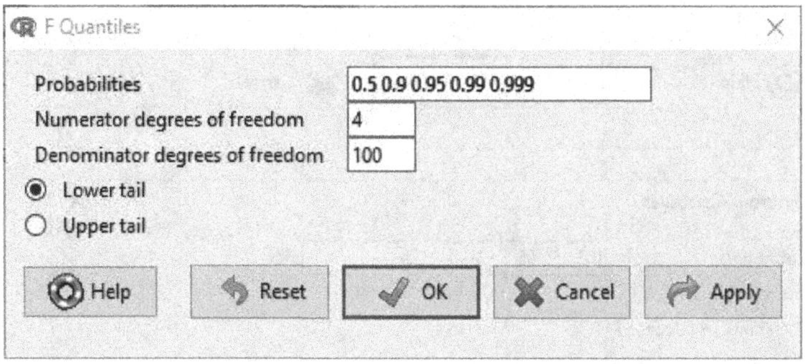

FIGURE 8.3: The *F Quantiles* dialog box.

The default selections in this dialog are similar to those in the *Normal Probabilities* dialog: Initially, the *Probabilities* text box in the dialog is empty, the *Mean* is set to 0, the *Standard deviation* is set to 1, and the *Lower tail* radio button is selected. I enter *Probabilities* .05, .025, .01, .005, set the *Mean* to 100 and *Standard deviation* to 15, and press the *Upper tail* radio button. Clicking *OK* produces the output

```
> qnorm(c(.05,.025,.01,.005), mean=100, sd=15, lower.tail=FALSE)
[1] 124.6728 129.3995 134.8952 138.6374
```

For example, in a normal distribution with mean $\mu = 100$ and standard deviation $\sigma = 15$, only $0.005 = 0.5\%$ of values exceed 138.6374.

As I mentioned, the probabilities and quantiles dialogs for other continuous distributional families have a similar format, although of course the content of each dialog reflects the parameters of the corresponding family. For example, Figure 8.3 shows the *F Quantiles* dialog, reached via *Distributions > Continuous distributions > F distribution > F quantiles*. Here, it's necessary to supply the probabilities for the desired quantiles, along with the numerator and denominator degrees of freedom, all of which are initially blank. I type 0.5 0.9 0.95 0.99 0.999 in the *Probabilities* box, and specify the numerator and denominator degrees of freedom as 4 and 100, respectively. Leaving the *Lower tail* radio button pressed and clicking *OK*, I get

```
> qf(c(0.5,0.9,0.95,0.99,0.999), df1=4, df2=100, lower.tail=TRUE)
[1] 0.8448915 2.0019385 2.4626149 3.5126841 5.0166504
```

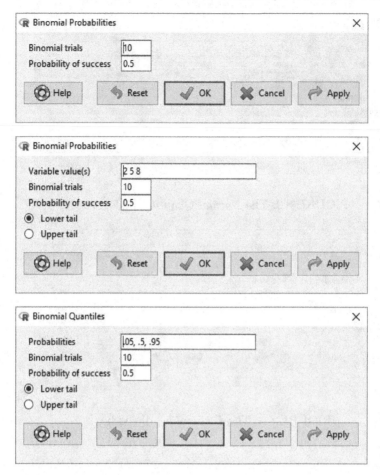

FIGURE 8.4: Binomial distributions dialogs: the probability-mass function (top), tail and cumulative probabilities (center), and quantiles (bottom). Both the probability-mass function and tail probabilities dialogs are named *Binomial Probabilities*.

8.1.2   Discrete Distributions

Dialogs for tail and cumulative probabilities and for quantiles of discrete distributions are largely similar to the probabilities and quantiles dialogs for continuous distributions. For discrete distributional families, however, there are both *tail probabilities* and *probabilities* dialogs, with the former printing tail and cumulative probabilities, and the latter printing a table of the *probability-mass function*. I will illustrate with the binomial family.

For example, choosing *Distributions > Discrete distributions > Binomial distribution > Binomial probabilities* brings up the dialog at the top of Figure 8.4, called *Binomial Probabilities*. Entering 10 *Binomial trials* (which has no default) and *Probability of success* 0.5 (which is the default value) returns a table of the entire probability distribution (probability-mass function), with the output displayed in Figure 8.5.

Similarly, selecting *Distributions > Discrete distributions > Binomial distribution > Binomial tail probabilities* produces the dialog in the middle of Figure 8.4 (also named *Binomial Probabilities*), where I enter the *Variable value(s)* 2 5 8, set *Binomial trials* to 10, set the *Probability of success* to 0.5, and select *Lower tail*:

```
> local({
+    .Table <- data.frame(Probability=dbinom(0:10, size=10, prob=0.5))
+    rownames(.Table) <- 0:10
+    print(.Table)
+ })
      Probability
0    0.0009765625
1    0.0097656250
2    0.0439453125
3    0.1171875000
4    0.2050781250
5    0.2460937500
6    0.2050781250
7    0.1171875000
8    0.0439453125
9    0.0097656250
10   0.0009765625
```

FIGURE 8.5: Output produced by the *Binomial Probabilities* probability-mass function dialog.

```
> pbinom(c(2,5,8), size=10, prob=0.5, lower.tail=TRUE)
[1] 0.0546875 0.6230469 0.9892578
```

Finally, selecting *Distributions > Discrete distributions > Binomial distribution > Binomial quantiles* leads to the dialog box at the bottom of Figure 8.4. I enter the *Probabilities* .05, .5, .95, set the *Binomial trials* to 10 (there are no defaults for these values), leave the *Probability of success* at the default, 0.5, leave the *Lower tail* radio button pressed, and click *OK*, obtaining[2]

```
> qbinom(c(.05,.5,.95), size=10, prob=0.5, lower.tail=TRUE)
[1] 2 5 8
```

8.2 Plotting Probability Distributions

The R Commander *Distributions* menu also includes items for drawing graphs of the *density function* and cumulative distribution function for continuous distributions, and graphs of the probability-mass function and CDF for discrete distributions. I'll illustrate with the continuous F distribution family and the discrete binomial family.

Distributions > Continuous distributions > F distribution > Plot F distribution produces the dialog in Figure 8.6, where I fill in 4 *Numerator degrees of freedom* and 100 *Denominator degrees of freedom*. Figure 8.7 shows the corresponding density and distribution function plots.[3]

Plotting discrete probability-mass functions and CDFs is essentially similar. For example, Figure 8.8 shows the dialog box for plotting the binomial distribution, setting the

---

[2]Perhaps curiously, for a discrete distribution like the binomial, the quantile and cumulative distribution functions aren't strictly inverses of each other: The quantile function $x = q(p)$ is defined as the *smallest* value x of the random variable X for which $\Pr(X \leq x) = \sum_{X \leq x} \Pr(X = x)$ is greater than or equal to

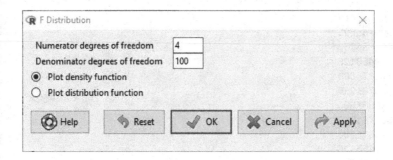

FIGURE 8.6: The *F Distribution* plotting dialog.

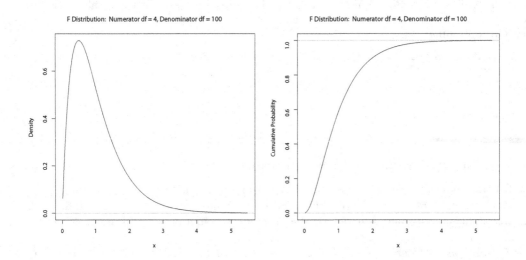

FIGURE 8.7: Density function (left) and cumulative distribution function (right) for the *F* distribution with 4 and 100 degrees of freedom.

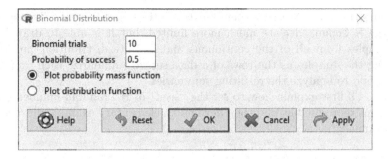

FIGURE 8.8: The *Binomial Distribution* plotting dialog.

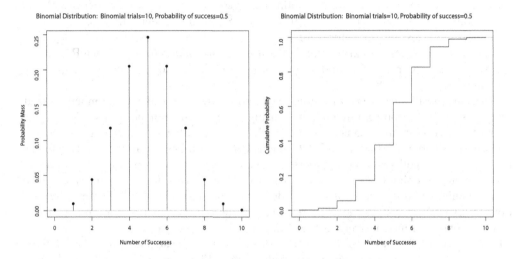

FIGURE 8.9: Probability-mass function (left) and cumulative distribution function (right) for the binomial distribution with probability of success $= 0.5$ and binomial trials $n = 10$.

Binomial trials to 10 and the *Probability of success* to 0.5. The corresponding plots of the probability-mass function and the CDF are shown in Figure 8.9.

8.3   Simple Random Simulations

Because it incorporates a general programming language with the ability to sample from a wide variety of distributions, along with extensive preprogrammed data-analytic capabilities, R is a powerful tool for constructing random "Monte Carlo" simulations. Simulation is useful both for teaching statistical ideas and for investigating properties of statistical

---

p, while the CDF function is defined as $P(x) = \sum_{X \leq x} \Pr(X = x)$. Thus, for example, for the binomial distribution with probability of success $= 0.5$ and $n = 10$ trials, $q(0.05) = 2$ but $P(2) = 0.0546875$.

[3]Although the density $p(f)$ is greater than 0 for *all* positive values f of an F random variable, the R Commander is smart enough to plot the density function and CDF only for values of F where the density isn't *effectively* 0: Of course, the R Commander can't draw the plot all the way up to $F = \infty$.

methods when analytic (i.e., direct mathematical) solutions are difficult.[4] The simulation capabilities of the R Commander are much more limited, but it is able to draw independent random samples from all of the continuous and discrete distributions in Table 8.1 (page 172), saving the samples as the rows of a data set, to summarize each sample using simple statistics, and to analyze the resulting summaries.

In this section, I'll first explain how to set the "seed" of R's random-number generator to make simulations reproducible, and then develop an elementary example illustrating the central limit theorem.

8.3.1   Setting the Seed of the R Pseudo-Random-Number Generator

Unless they have specialized hardware, computers can't generate truly random numbers. Think about it: Computers execute deterministic programs; run a program twice with exactly the same inputs and you get the same results.[5] To circumvent this problem, clever programmers have devised procedures for generating numbers that behave *as if* they were random; these procedures are called *pseudo-random-number generators*, of which R incorporates several. The R Commander (and almost all users of R) employs the default R pseudo-random-number generator.[6]

An advantage of using *pseudo*-random (as opposed to *truly* random) numbers is that you can make a simulation reproducible by reusing a particular sequence of pseudo-random numbers. R's random-number generator starts with a value called a *seed*, which may be any integer between about -2 billion and $+2$ billion. Use the same seed and you generate the same sequence of pseudo-random numbers. By setting the seed to a known value, you can therefore replicate a simulation exactly. This procedure insures, for example, that if you run your simulation using the R Markdown document produced by the R Commander (see Section 3.6), you'll always get the same results, and these results will be the same as those you initially obtained interactively in the R Commander *Output* pane.

The most convenient way to set the random seed is via *Distributions > Set random number generator seed*, which produces the dialog box in Figure 8.10. The dialog includes a slider that ranges from 1 to 100,000, and is initially set to a pseudo-random value based on the time of day; the initial value in Figure 8.10 is 19,878. There's no reason not to accept this value by clicking *OK*: What's important is that the seed is a *known value*, not what that value is.[7]

8.3.2   A Simple Simulation Illustrating the Central Limit Theorem

You almost surely have encountered—or will encounter—the *central limit theorem* in a basic statistics course: Suppose that X is distributed in the population with mean μ and variance σ^2. Then, almost regardless of the shape of the distribution of X, the sampling distribution of means $\bar{X}$ for repeated independent random samples of size n drawn from the population will be approximately normal, with mean $E(\bar{X}) = \mu$ and variance $V(\bar{X}) \doteq \sigma^2/n$, and this approximation improves as n grows. That is, as $n \to \infty$, $\bar{X} \to N(\mu, \sigma^2/n)$. The values of

---

[4]Simulation also plays an increasingly important role in data analysis. For example, the *bootstrap* (see, e.g., Efron and Tibshirani, 1993) is a method of statistical inference based on randomly resampling from the data. Similarly, modern Bayesian statistical inference (as described, e.g., by Gelman et al., 2013) is heavily dependent on simulation.

[5]This might not literally be true because a program is executed in an environment; there might be a hardware failure, for example.

[6]The details need not concern us, but if you're interested, type the command `?Random` at the prompt in the R console.

[7]If, however, you're trying to reproduce an earlier session that used a particular known seed, then you must set the seed for the current session to the known value. For example, if you want to reproduce my results in this section exactly, set the seed to 19,878 (and duplicate my sequence of operations exactly).

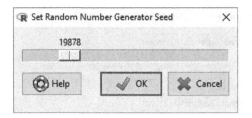

FIGURE 8.10: The *Set Random Number Generator Seed* dialog; the initial value 19878 was itself pseudo-randomly generated.

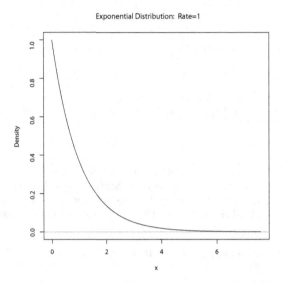

FIGURE 8.11: Density function of the highly positively skewed exponential distribution with rate parameter $= 1$.

$E(\bar{X})$ and $V(\bar{X})$ are exact for any n; it's the normal distribution of $\bar{X}$ that's an *asymptotic approximation*.

I'll illustrate the central limit theorem by drawing 10,000 repeated samples, for each of several sample sizes, from an exponential population with *rate parameter* 1. The mean and standard deviation of an exponential random variable are equal to the inverse of the rate parameter, and so $\mu = \sigma = 1^{-1} = 1$. The density function for this distribution is highly positively skewed, as shown in Figure 8.11 (produced via *Distributions > Continuous distributions > Exponential distribution > Plot exponential distribution*).

Selecting *Distributions > Continuous distributions > Exponential distribution > Sample from exponential distribution* brings up the dialog box in Figure 8.12, which shows the default values for the dialog. I change the name of the data set to `ExponentialSamples2`, the *Number of samples* to 10000, and the `Number of observations` in each sample (i.e., n) to 2. I leave the *Sample means* box checked to compute the mean $\bar{x}$ of the $n = 2$ observations in each sample—that is, one mean for each of the 10,000 samples. The resulting data set, with 10,000 rows and three columns (the two observations for each sample and their mean) becomes the active data set in the R Commander.

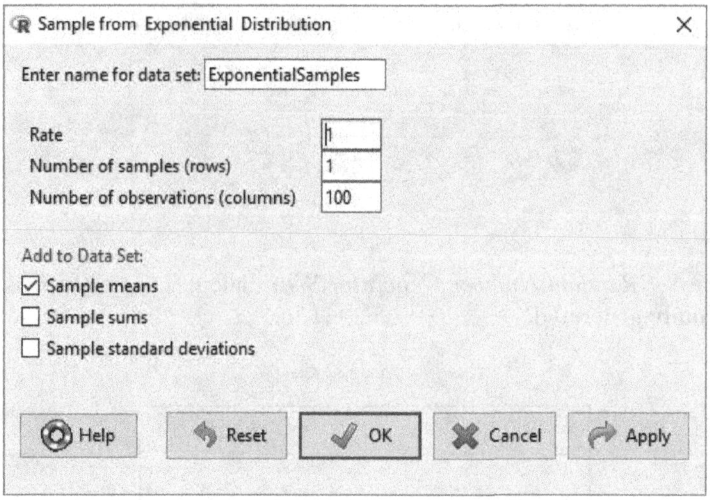

FIGURE 8.12: The initial state of the *Sample from Exponential Distribution* dialog, showing default selections.

Because the data set has samples as rows, the sample mean is a *variable* in the data set, and I can examine its distribution. Selecting *Statistics > Summaries > Numerical Summaries* brings up the *Numerical Summaries* dialog box (see Section 5.1), in which I compute the mean and standard deviation of the **mean** (i.e., of the 10,000 sample means), obtaining

```
> numSummary(ExponentialSamples2[,"mean"], statistics=c("mean", "sd"),
+    quantiles=c(0,.25,.5,.75,1))
      mean        sd        n
 0.9969254 0.7027704 10000
```

These values agree nicely with statistical theory: $E(\bar{X}) = \mu = 1$ and $\mathrm{SD}(\bar{X}) = \sigma/\sqrt{n} = 1/\sqrt{2} \approx 0.707$. The slight slippage is due to the fact that I have 10,000 rather than an infinite number of samples.

A histogram of the 10,000 sample means, produced via *Graphs > Histogram* (see Section 5.3.1), appears at the upper left of Figure 8.13: Clearly, when $n = 2$, the distribution of the sample means is itself positively skewed, although not quite as skewed as the exponential population from which I sampled.

I repeated the simulation successively for $n = 5$, 25, and 100, producing a histogram of sample means for each data set of 10,000 samples. The results also appear in Figure 8.13. In comparing the histograms, be careful to notice that the scaling of both the horizontal and vertical axes (i.e., the units per cm) changes. Taking the change in scaling of the horizontal axis into account, it is apparent that, as the sample size grows, the sample means become less variable. It's also clear from the shape of the histograms that, as the sample size grows, the distribution of sample means becomes more symmetric and generally more like a normal distribution.[8]

Simulations in the R Commander are a bit more powerful than they first appear, because the data manipulation tools provided by the R Commander (described in Chapter 4) allow you to merge data sets that sample from different distributions and permit you to compute

---

[8]If you're familiar with theoretical quantile-comparison plots, a more effective way of comparing the distributions of sample means to the normal distribution would be via *Graphs > Quantile-comparison plot*.

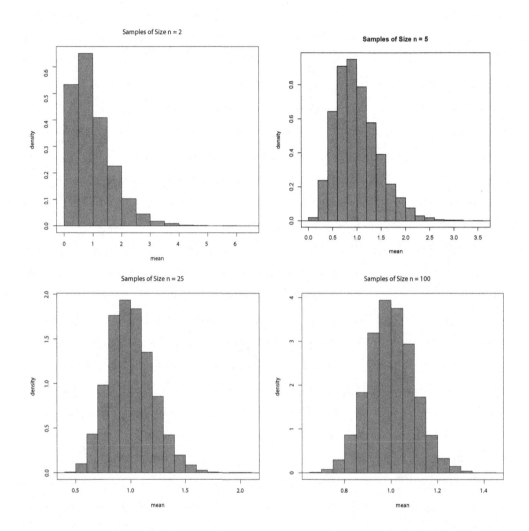

FIGURE 8.13: Histograms of sample means of size n computed for 10,000 samples drawn from an exponential population with rate parameter 1, for each of $n = 2$, 5, 25, and 100 (from upper-left to lower-right). The vertical axis of each histogram is density scaled so that the areas of the bars in each histogram sum to 1. The scaling of both axes (in units/cm) changes from graph to graph.

arbitrary expressions based on the values in each row of a simulated data set. Nevertheless, it is much more natural to use the command line interface to R to develop simulations, leveraging the programmability of R (see Section 1.4).

9

Using *R Commander* Plug-in Packages

The capabilities of the R Commander are substantially augmented by the many *plug-in packages* for it that are available on CRAN. I provided for plug-in packages relatively early in the development of the R Commander. Although the details need not concern us,[1] R Commander plug-ins integrate with the R Commander menus, producing standard dialog boxes, accessing and possibly modifying the active data set, creating commands in the R Commander *R Script* and *R Markdown* tabs, and printing results in the R Commander *Output* pane. Plug-ins can also add new classes of statistical models, manipulable via the R Commander *Models* menu.[2]

This chapter explains how to install plug-in packages, and illustrates the application of R Commander plug-ins by using the **RcmdrPlugin.TeachingDemos** package and the **RcmdrPlugin.survival** package as examples.

9.1 Acquiring and Loading Plug-Ins

As I write this, there are about 40 R Commander plug-in packages on CRAN. The recommended naming convention for R Commander plug-ins is **RcmdrPlugin.***name*, and consequently you can discover what's available by going to the CRAN web page that lists packages by name (https://cran.r-project.org/web/packages/available_packages_by_name.html), and searching for the text "RcmdrPlugin."[3] Each CRAN package has a description page—just click on the link for the package—that provides information about the package.

Because R Commander plug-ins are standard CRAN packages, they can be installed in the usual manner. For example, to install the two plug-in packages discussed in this chapter, issue the following command at the > prompt in the R console:[4]

```
install.packages(c("RcmdrPlugin.TeachingDemos", "RcmdrPlugin.survival"))
```

There are two ways to load R Commander plug-in packages:

1. Select *Tools > Load Rcmdr plugin(s)* from the R Commander menus, producing the dialog box in Figure 9.1, which lists the plug-in packages installed in your

---

[1]You'll find a manual for R Commander plug-in package writers (as opposed to *users*) on the web site for this book.

[2]All plug-in packages should work properly with the R Commander, but need not be compatible with *each other*. Plug-ins are given wide latitude to modify the R Commander interface; so one plug-in package, for example, might remove a menu to which another plug-in tries to add a menu item, causing an error. If you experience a problem with an R Commander plug-in, you should normally write first to the author of the package, not to me—unless I *am* the plug-in package author, as is the case for the two plug-ins discussed in this chapter.

[3]A few R Commander plug-ins don't follow this naming convention. You can discover these nonconforming plug-ins by navigating to the **Rcmdr** package description page on CRAN and looking at the package's "reverse dependencies."

[4]Alternatively use the menus in the *R Console*: on Windows, *Packages > Install package(s)*; on Mac OS X, *Packages & Data > Package Installer*.

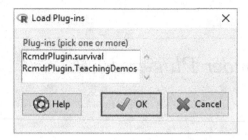

FIGURE 9.1: The *Load Plug-ins* dialog box. Two plug-in packages, **RcmdrPlugin.survival** and **RcmdrPlugin.TeachingDemos** are installed.

package library. Simply pick the plug-in (or plug-ins) that you want to load. After giving you a chance to save your work, the R Commander will restart, loading the selected plug-in and beginning a fresh session.

2. Some plug-ins are *self-starting* and can be loaded directly by a `library` command issued in the R console. For example, the plug-ins discussed in this chapter are both self-starting. Thus, the command `library(RcmdrPlugin.survival)` loads the **RcmdrPlugin.survival** package along with the **Rcmdr** package, and starts up the R Commander GUI.

9.2   Using the RcmdrPlugin.TeachingDemos Package

I created the **RcmdrPlugin.TeachingDemos** package (see Fox, 2007) primarily to demonstrate how to write an R Commander plug-in. The plug-in integrates some of the demonstrations in Greg Snow's **TeachingDemos** package (Snow, 2013) into the R Commander, and may be of interest to students and teachers of basic statistics.

As explained in the preceding section, entering the command `library(RcmdrPlugin.TeachingDemos)` in a fresh R session loads both the **Rcmdr** and the **RcmdrPlugin.TeachingDemos** packages. The plug-in simply adds a new top-level *Demos* menu (the contents of which are displayed in Figure 9.2) to the R Commander menu bar.

The purpose of the various items in the *Demos* menu is reasonably clear from the names of the items. Two of the items, *Central limit theorem* and *Confidence interval for the mean*,

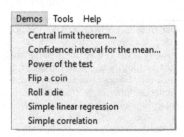

FIGURE 9.2: The *Demos* top-level menu and menu items added by the **RcmdrPlugin.TeachingDemos** package.

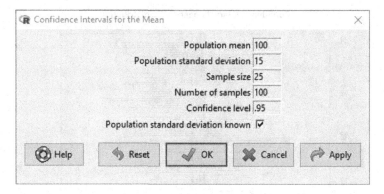

FIGURE 9.3: *Confidence Intervals for the Mean* dialog box provided by the **RcmdrPlugin.TeachingDemos** plug-in package.

lead to dialogs provided by the plug-in. *Power of the test, Simple linear regression,* and *Simple correlation* bring up interactive demonstrations supplied directly by the **TeachingDemos** package. *Flip a coin* and *Roll a die* produce 3D animations.

I'll illustrate with the confidence interval demo, but please feel free to experiment with the other demos. Selecting *Demos > Confidence interval for the mean* produces the dialog in Figure 9.3. All of the selections in the dialog are defaults except for the *Number of samples,* which I changed from 50 to 100.

Thus, I'll draw 100 repeated simulated samples, each of size $n = 25$, from a normal population with mean $\mu = 100$ and standard deviation $\sigma = 15$. A 95% confidence interval will be constructed around each sample mean, assuming that the population standard deviation is known:[5]

$$\bar{x} \pm 1.96\sigma/\sqrt{n} = \bar{x} \pm 1.96 \times 15/\sqrt{25} = \bar{x} \pm 5.88$$

Pressing the *Apply* button in the dialog produces the graph in Figure 9.4.[6] The black vertical line in the graph represents the fixed value of $\mu = 100$, and the two red lines are drawn at $\mu \pm 5.88$, the central interval within which 95% of sample means will fall with infinitely repeated sampling. The original graph, which is in color, appears in the insert at the center of the book. In the color version of the figure, short black vertical lines mark the 100 individual sample means, the $\bar{x}$s, while the horizontal lines represent the 100 confidence intervals, of equal width 2×5.88, centered at the values of $\bar{x}$. The black confidence intervals include $\mu = 100$; the magenta confidence intervals miss low; and the cyan confidence intervals miss high.

As you can see, in my simulation there are 93 "hits" and 7 "misses." At 95% confidence, I *expect* 95 hits and 5 misses in 100 samples, but of course these are averages, and random variation is to be anticipated. Try pressing the *Apply* button a few times to get a sense of how the numbers of hits and misses vary from one simulation of 100 samples to the next.

---

[5]If you're not already familiar with this result, you'll surely encounter it in your basic statistics course.

[6]Because the samples are drawn randomly, and because I haven't bothered to set the random seed to a known value (see Section 8.3 on simulation), you won't get exactly the same results as mine if you repeat this demonstration, but your results should be roughly similar.

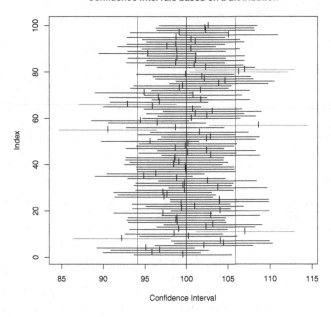

FIGURE 9.4: Ninety-five percent confidence intervals based on the means $\bar{x}$ of 100 samples, each of size $n = 25$, drawn from a normal population with mean $\mu = 100$ and (known) standard deviation $\sigma = 15$. A color version of this figure appears in the insert at the center of the book.

9.3 Survival Analysis with the RcmdrPlugin.survival Package*

Survival analysis is concerned with the timing of events—the event of death in many biomedical applications, hence the terminology—and is widely employed under a number of synonymous terms (*event-history analysis, duration analysis, failure-time analysis*) in a variety of disciplines. The **survival** package (Therneau, 2015; Therneau and Grambsch, 2000) is state-of-the-art survival-analysis software, and is part of the standard R distribution. The **RcmdrPlugin.survival** package adds a graphical interface for many of the methods in the **survival** package, including survival-function estimation, parametric survival regression, and Cox proportional-hazards regression.

I'm using the **RcmdrPlugin.survival** package here primarily to illustrate a more ambitious R Commander plug-in: As opposed to the **RcmdrPlugin.TeachingDemos** package (described in the preceding section), which simply adds a new top-level menu to the R Commander, the **RcmdrPlugin.survival** plug-in is tightly integrated into the R Commander, adding several sub-menus and menu items to existing R Commander menus, and defining two new classes of statistical models, for parametric and Cox survival regression. It's not my object, however, to cover all of the capabilities of the **RcmdrPlugin.survival** package; for a more detailed discussion of the package, see Fox and Carvalho (2012).

The **RcmdrPlugin.survival** additions to the R Commander menus are shown in Figures 9.5 and 9.6.

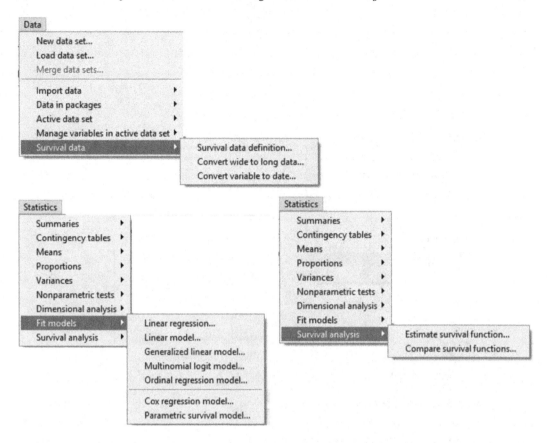

FIGURE 9.5: **RcmdrPlugin.survival** additions to the R Commander *Data* and *Statistics* menus.

- The *Data* menu gains a *Survival data* sub-menu, with items for defining survival data (e.g., identifying time and event variables), for converting data sets from "wide" format to "long" format,[7] and for handling dates.

- Cox and parametric survival regression models are added to the *Statistics > Fit models* menu.

- There is a new *Survival analysis* sub-menu under the *Statistics* menu, with menu items for estimating and comparing survival functions.

- Once a survival regression model is fit, it may be manipulated with some of the standard items in the *Models* menu. In addition, there are new menu items for testing proportional hazards in the Cox model, and for drawing diagnostic and interpretative graphs for Cox models and (to a lesser extent) parametric survival regression models.

---

[7]In wide format, time-varying variables for each individual appear as successive values in a single row of the data set, while in long format, there are separate rows representing the various time intervals for each individual.

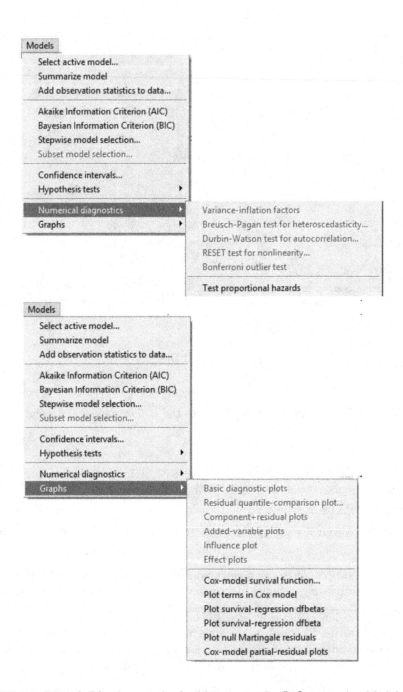

FIGURE 9.6: **RcmdrPlugin.survival** additions to the R Commander *Models* menu.

9.3.1   Survival Analysis Examples Based on Criminal Recidivism Data

The `Rossi` data set in the **RcmdrPlugin.survival** package is drawn from a study of criminal recidivism by Rossi et al. (1980). The data are employed extensively by Allison (2014) in a fine monograph on survival analysis. I'll duplicate some of Allison's results here using the R Commander with **RcmdrPlugin.survival**.

Rossi et al.'s data pertain to 432 male convicts who were released from Maryland state prisons in the 1970s, and who were followed up on a weekly basis for one year subsequent to their release. In a randomized experiment, half of the convicts were given financial aid upon release and half didn't receive aid. The goal of the study was to determine whether receiving financial aid reduces the risk of rearrest.

The variables in the `Rossi` data set are as follows (with names corresponding to those employed by Allison, 2014):

- `week` The week of first arrest after release or the censoring time; all censored cases are censored at 52 weeks, when follow-up ceased.

- `arrest` An event indicator, coded 1 if the former convict was rearrested and 0 if censored.

- `fin` Whether or not the former convict received financial aid, a factor coded `yes` or `no`.

- `age` The convict's age in years at time of release.

- `race` The convict's race, a factor coded `black` or `other`.

- `wexp` Whether or not the convict had full-time work experience prior to incarceration, a factor coded `yes` or `no`.

- `mar` The convict's maritial status at the time of release, a factor coded `married` or `not married`.

- `paro` Whether or not the convict was released on parole, a factor coded `yes` or `no`.

- `prio` The convict's number of convictions prior to his most recent incarceration.

- `educ` The convict's educational level, numerically coded as 2 (less than 6th grade), 3 (7th to 9th grade), 4 (10th or 11th grade), 5 (12th grade), or 6 (some post-secondary).

- `emp1–emp52` Employment status in each week of the study, factors coded `yes` or `no` according to whether or not the released convict was employed during the corresponding week. Once a convict is rearrested, subsequent values of `empx` are `NA` (missing).[8]

After loading the **Rcmdr** and **RcmdrPlugin.survival** packages in a fresh session with the command `library(RcmdrPlugin.survival)`,[9] I read the `Rossi` data into the R Commander in the usual manner, via *Data > Data in packages > Read data from an attached package* (see Section 4.2.4). Then *Data > Survival data > Survival data definition* brings up the dialog box in Figure 9.7, in which I select `week` as the *Time* variable and `arrest` as the *Event indicator*, leaving the other selections in the dialog at their defaults. It's not necessary to define time and event variables in this manner, but it's convenient to do so, because these variables will be preselected in subsequent survival analysis dialogs.

---

[8]The data, therefore, are in wide format, with all of the values of the time-varying employment covariate appearing in the same row for each released convict.

[9]It would also work to continue the previous session in this chapter, loading the **RcmdrPlugin.survival** plug-in via the R Commander menus, *Tools > Load Rcmdr Plugin(s)*.

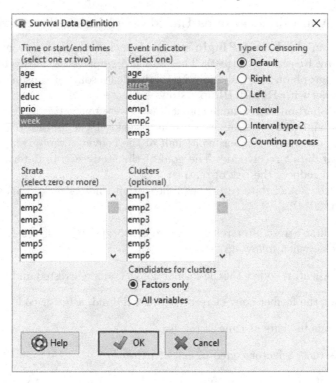

FIGURE 9.7: *Survival Data Definition* dialog for the `Rossi` data.

Choosing *Statistics > Survival analysis > Estimate survival function* produces the dialog shown in Figure 9.8. As mentioned, `week` and `arrest` are preselected as the *Time* and *Event* variables. Scrolling down in the variable list, I pick `fin` for *Strata*, to estimate separate survival functions for those receiving and not receiving financial aid; the default is not to define strata. I leave all of the selections in the *Options* tab at their defaults. The resulting graph appears in Figure 9.9; the dialog also produces some printed output (which is not shown).

Here, "survival" represents staying out of jail, and, apparently, the former convicts who received financial aid upon release were slightly more likely to stay out of jail than those who didn't receive aid. To test this difference, I choose *Statistics > Survival analysis > Compare survival functions*, bringing up the dialog in Figure 9.10. I once again select `fin` in the *Strata* list box and accept all of the other defaults. In particular, *rho = 0* corresponds to the commonly employed *log-rank* or *Mantel–Haenszel test*. The output is shown in Figure 9.11. The two survival functions are nearly but not quite statistically significantly different at the conventional $\alpha = .05$ level, although arguably a one-sided test, halving the reported *p*-value, would be sensible here.

Perhaps the comparison between the two financial-aid groups can be sharpened by adding covariates to the analysis. I pursue that idea by fitting a Cox proportional-hazards regression to the data, via the menu selection *Statistics > Fit models > Cox regression model*, bringing up the dialog box in Figure 9.12. I retain the default selections in the *Data* tab, and enter the various covariates, including `fin` (and with the exception of the time-varying employment covariate), into the model. The specification of the right-hand side of the Cox-

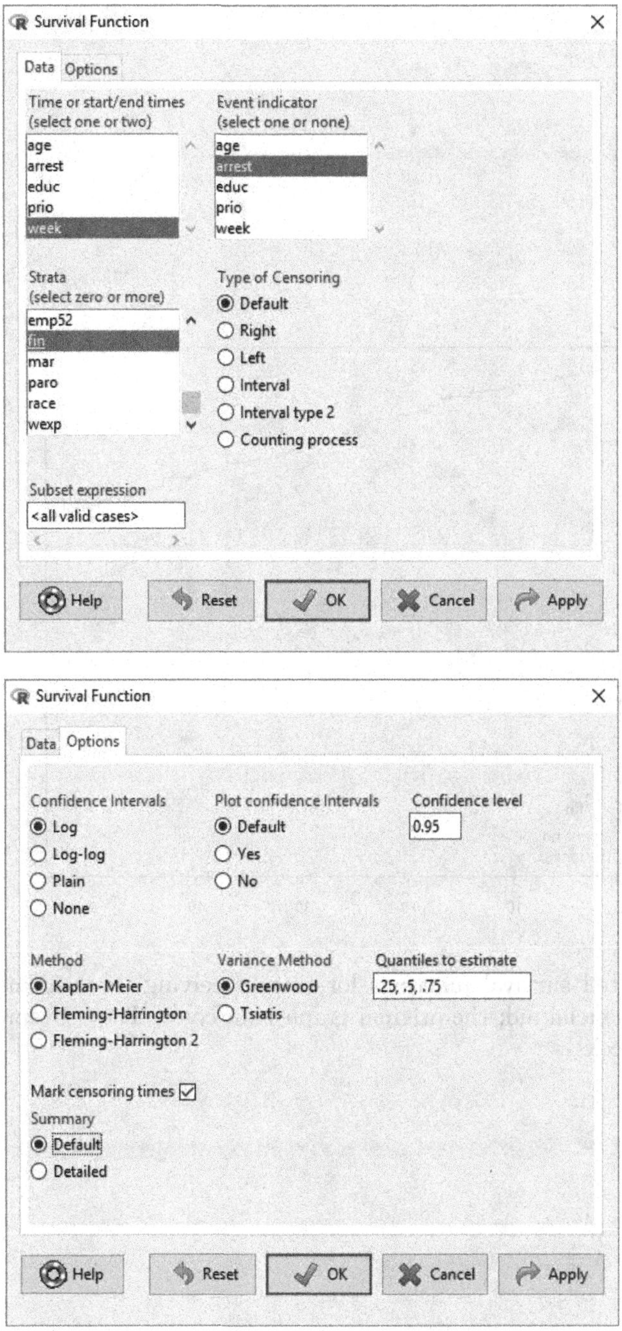

FIGURE 9.8: The *Survival Function* dialog, *Data* and *Options* tabs.

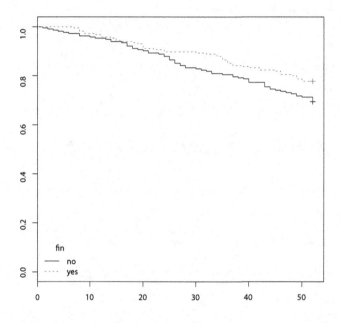

FIGURE 9.9: Estimated survival functions for those receiving (broken line) and not receiving (solid line) financial aid; the original graph is in color. The +s represent censored observations (at 52 weeks).

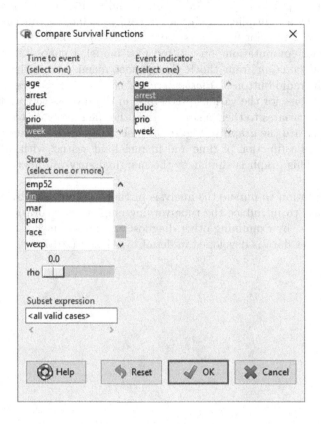

FIGURE 9.10: The *Compare Survival Functions* dialog.

```
> survdiff(Surv(week,arrest) ~ fin, rho=0, data=Rossi)
Call:
survdiff(formula = Surv(week, arrest) ~ fin, data = Rossi, rho = 0)

          N Observed Expected (O-E)^2/E (O-E)^2/V
fin=no   216       66     55.6      1.96      3.84
fin=yes  216       48     58.4      1.86      3.84

 Chisq= 3.8  on 1 degrees of freedom, p= 0.0501
```

FIGURE 9.11: Test of the difference in survival functions for those receiving and not receiving financial aid.

model formula is essentially the same as for a linear model (discussed in Section 7.2).[10] Clicking *OK* results in the output in Figure 9.13. Financial aid is therefore estimated to reduce the hazard of rearrest by a multiplicative factor of $\exp(b_{\text{fin}}) = 0.6979$, but this coefficient is just barely statistically significant by a one-sided Wald test, $p = 0.06079/2 = 0.03039$.

To illustrate further computations on a fitted Cox model, I choose *Models > Graphs > Cox-model survival function* from the R Commander menus, producing the dialog in Figure 9.14. I press the radio button to *Plot at specified values of predictors*, move the slider to *2* rows, and enter values for the various predictors, in the process setting `fin` to each of `no` and `yes`, numeric covariates to their medians, and other factor covariates to their modal levels. These choices create the graph in Figure 9.15, showing the estimated probability of remaining out of jail as a function of time and financial-aid status, with other covariates set to typical values. This graph is similar to the marginal survival-function estimates in Figure 9.9 (page 192).

I'll resist the temptation to pursue the analysis further, for example, by converting the data set to long format to introduce the time-varying employment covariate, by checking for proportional hazards, by examining other diagnostics, and so on. For those interested, the analysis of the `Rossi` data is developed in detail by Allison (2014).

---

[10]There is no intercept, however, in a Cox model; in effect, the role of the intercept is played by the *baseline hazard function*. As usual, it's not my goal to explicate statistical methods but rather to illustrate their application in the R Commander.

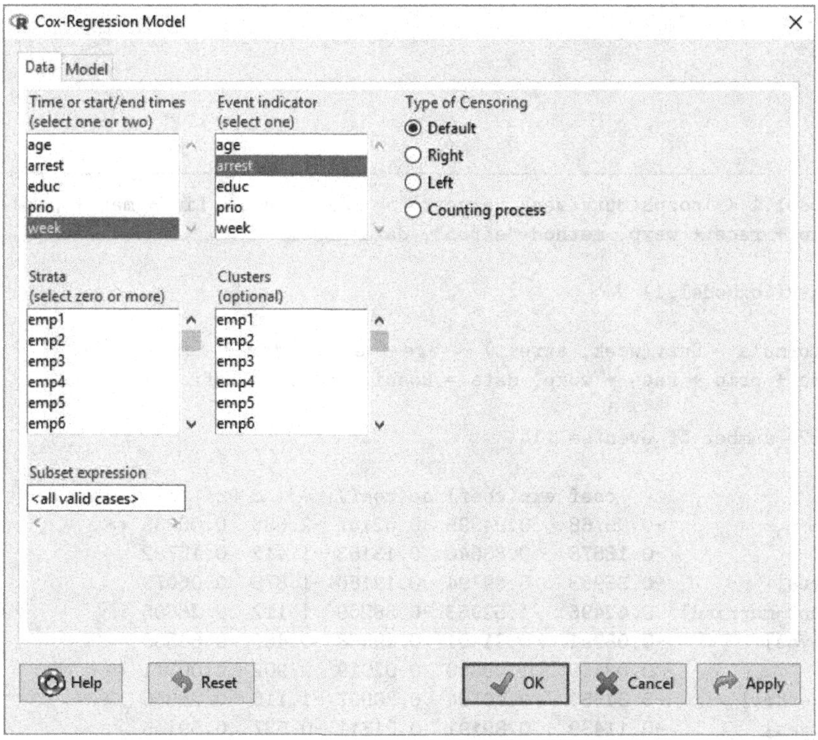

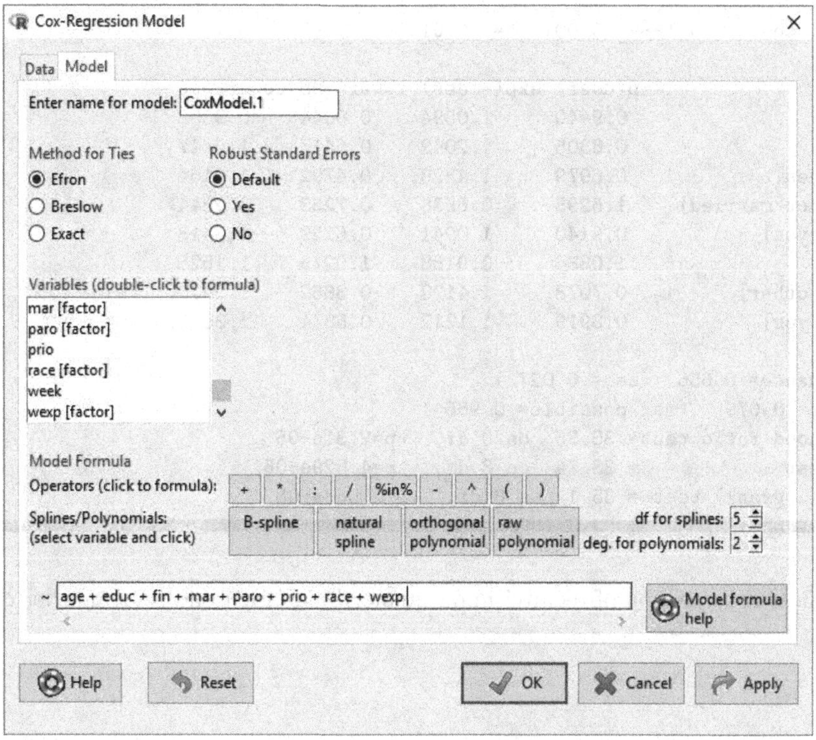

FIGURE 9.12: The *Cox-Regression Model* dialog, *Data* and *Model* tabs.

```
> CoxModel.1 <- coxph(Surv(week, arrest) ~ age + educ + fin + mar + paro +
+    prio + race + wexp, method="efron", data=Rossi)

> summary(CoxModel.1)
Call:
coxph(formula = Surv(week, arrest) ~ age + educ + fin + mar +
    paro + prio + race + wexp, data = Rossi, method = "efron")

  n= 432, number of events= 114

                     coef exp(coef) se(coef)      z Pr(>|z|)
age              -0.05768   0.94395  0.02187 -2.638  0.00835 **
educ             -0.18578   0.83046  0.13153 -1.412  0.15782
fin[T.yes]       -0.35963   0.69794  0.19180 -1.875  0.06079 .
mar[T.not married] 0.42496  1.52953  0.38209  1.112  0.26605
paro[T.yes]      -0.08991   0.91401  0.19568 -0.459  0.64589
prio              0.08469   1.08838  0.02919  2.902  0.00371 **
race[T.other]    -0.34554   0.70784  0.30907 -1.118  0.26356
wexp[T.yes]      -0.11439   0.89191  0.21311 -0.537  0.59145
---
Signif. codes:  0 '***' 0.001 '**' 0.01 '*' 0.05 '.' 0.1 ' ' 1

                   exp(coef) exp(-coef) lower .95 upper .95
age                   0.9440     1.0594    0.9044    0.9853
educ                  0.8305     1.2042    0.6417    1.0747
fin[T.yes]            0.6979     1.4328    0.4792    1.0164
mar[T.not married]    1.5295     0.6538    0.7233    3.2344
paro[T.yes]           0.9140     1.0941    0.6229    1.3413
prio                  1.0884     0.9188    1.0279    1.1525
race[T.other]         0.7078     1.4128    0.3862    1.2972
wexp[T.yes]           0.8919     1.1212    0.5874    1.3543

Concordance= 0.656  (se = 0.027 )
Rsquare= 0.079   (max possible= 0.956 )
Likelihood ratio test= 35.35  on 8 df,   p=2.31e-05
Wald test          = 33.74  on 8 df,   p=4.529e-05
Score (logrank) test = 35.1  on 8 df,   p=2.568e-05
```

FIGURE 9.13: Cox regression of recidivism on financial aid (fin) and several other covariates.

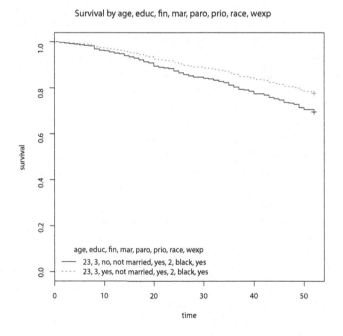

FIGURE 9.14: The *Plot Cox-Model Survival Functions* dialog, setting `fin` to `no` and `yes` and other covariates to typical values.

FIGURE 9.15: Estimated survival functions for those receiving (broken line) and not receiving (solid line) financial aid, based on the Cox model fit to the `Rossi` data; covariates other than `fin` are held to typical values.

Appendix A

Guide to the *R Commander* Menus

The expanded R Commander menus are shown on subsequent pages in Figures A.1 through A.9, as described in Table A.1.[1] Most of the menus are discussed in this book, as also indicated in the table.[2] If any optional auxiliary software (Pandoc or LaTeX) is missing, then an additional *Install auxiliary software* item appears under the *Tools* menu (Figure A.9).[3] Finally, on Mac OS X systems, a *Manage Mac OS X app nap for R.app* item also appears in the *Tools* menu.[4]

TABLE A.1: The R Commander menus.

Figure(s)	Menu(s)	Discussed in
A.1	*File* and *Edit*	Secs. 3.6–3.8
A.2 and A.3	*Data*	Secs. 3,3, 3.4, Ch. 4, Sec. 9.3
A.4 and A.5	*Statistics*	Secs. 3.4, 3.5, 5.1, 5.2, Ch. 6, Secs. 7.1–7.5, 9.3
A.6	*Graphs*	Secs. 3.4, 5.3, 5.4
A.7	*Models*	Secs. 7.5–7.9, 9.3
A.8	*Distributions*	Ch. 8
A.9	*Tools* and *Help*	2.3.3, 2.5, 3.9, 4.2.4, Ch. 9

---

[1]These menus are current as of version 2.2-4 of the **Rcmdr** package. In the interest of brevity, the *Distributions* menu isn't fully expanded, showing sub-menus and menu items only for one continuous distribution (the normal distribution) and one discrete distribution (the binomial distribution).

[2]An exception is the *Statistics > Dimensional analysis* menu, with menu items for scale reliability, principal component analysis, and exploratory and confirmatory factor analysis, and a sub-menu for cluster analysis.

[3]See the information about installing optional auxiliary software in Section 2.5.

[4]See Section 2.3.3 for a discussion of app nap in Mac OS X.

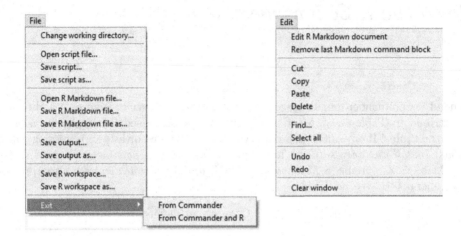

FIGURE A.1: The R Commander *File* and *Edit* menus.

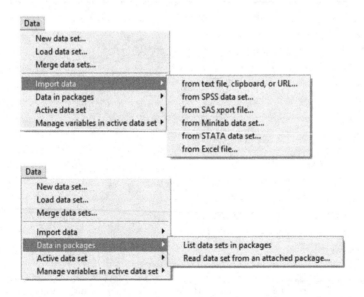

FIGURE A.2: The R Commander *Data > Import data* and *Data > Data in packages* menus.

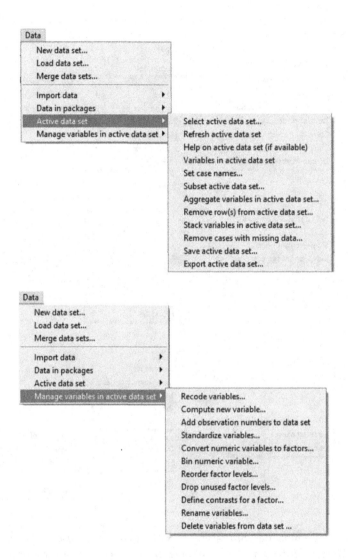

FIGURE A.3: The R Commander *Data > Active data set* and *Data > Manage variables in active data set* menus.

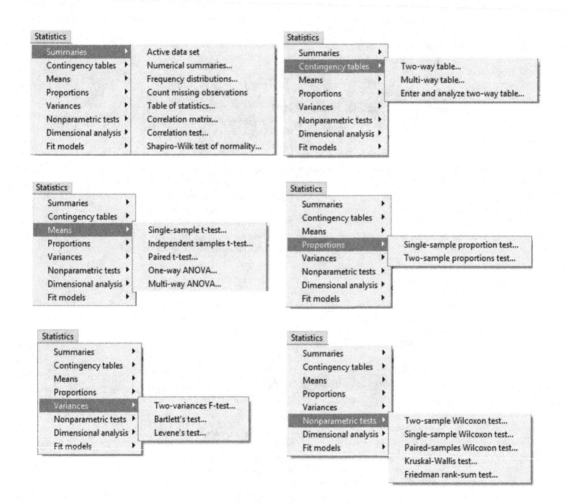

FIGURE A.4: The R Commander *Statistics > Summaries, Statistics > Contingency tables, Statistics > Means, Statistics > Proportions, Statistics > Variances*, and *Statistics > Nonparametric tests* menus.

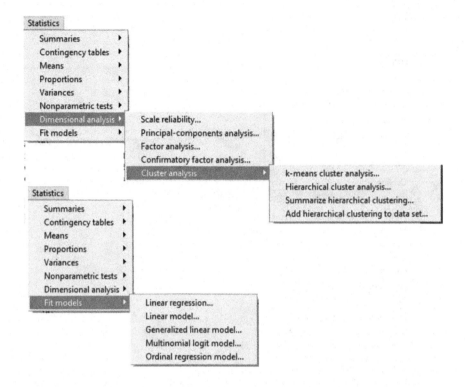

FIGURE A.5: The R Commander *Statistics > Dimensional analysis* and *Statistics > Fit models* menus.

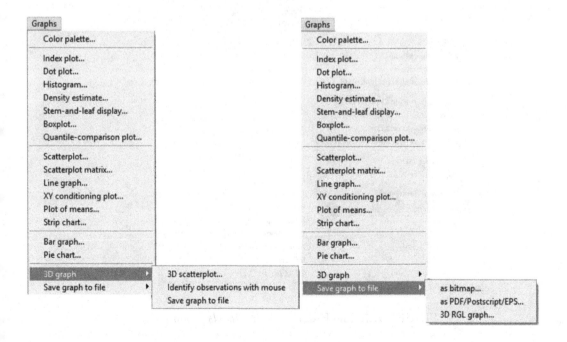

FIGURE A.6: The R Commander *Graphs* menu.

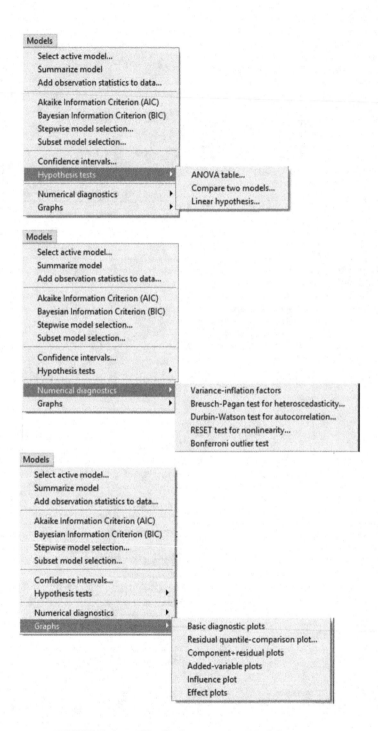

FIGURE A.7: The R Commander *Models* menu.

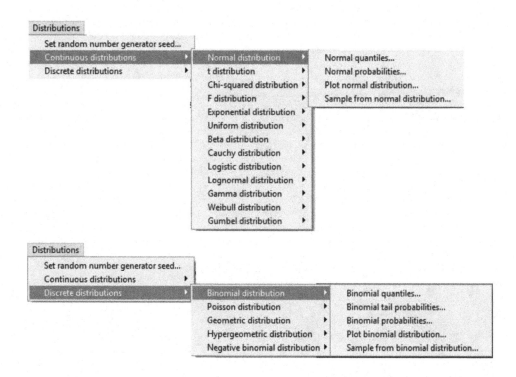

FIGURE A.8: The R Commander *Distributions* menu, showing the *Normal distribution* and *Binomial distribution* sub-menus.

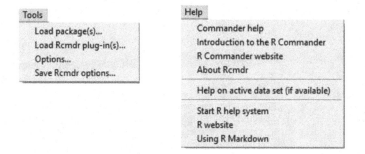

FIGURE A.9: The R Commander *Tools* and *Help* menus.

References

D. Adler and D. Murdoch. *rgl: 3D visualization using OpenGL*, 2015. URL http://CRAN.R-project.org/package=rgl. R package version 0.95.1367.

N. E. Adler. Impact of prior sets given experimenters and subjects on the experimenter expectancy effect. *Sociometry*, 36:113–126, 1973.

J. J. Allaire, J. Cheng, Y. Xie, J. McPherson, W. Chang, J. Allen, H. Wickham, A. Atkins, and R. Hyndman. *rmarkdown: Dynamic documents for R*, 2015a. URL http://CRAN.R-project.org/package=rmarkdown. R package version 0.8.1.

J. J. Allaire, J. Horner, V. Marti, and N. Porte. *markdown: 'Markdown' rendering for R*, 2015b. URL http://CRAN.R-project.org/package=markdown. R package version 0.7.7.

P. D. Allison. *Event History and Survival Analysis*. Sage, Thousand Oaks, CA, second edition, 2014.

R. A. Becker, J. M. Chambers, and A. R. Wilks. *The New S Language: A Programming Environment for Data Analysis and Graphics*. Wadsworth, Pacific Grove, CA, 1988.

T. S. Breusch and A. R. Pagan. A simple test for heteroscedasticity and random coefficient variation. *Econometrica*, 47:1287–1294, 1979.

A. Campbell, P. E. Converse, W. E. Miller, and D. E. Stokes. *The American Voter*. Wiley, New York, 1960.

J. M. Chambers and T. J. Hastie, editors. *Statistical Models in S*. Wadsworth, Pacific Grove, CA, 1992.

R. D. Cook and S. Weisberg. *Residuals and Influence in Regression*. Chapman and Hall, New York, 1982.

R. D. Cook and S. Weisberg. Diagnostics for heteroscedasticity in regression. *Biometrika*, 70:1–10, 1983.

M. Cowles and C. Davis. The subject matter of psychology: Volunteers. *British Journal of Social Psychology*, 26:97–102, 1987.

O. D. Duncan. A socioeconomic index for all occupations. In A. J. Reiss Jr., editor, *Occupations and Social Status*, pages 109–138. Free Press, New York, 1961.

J. Durbin and G. S. Watson. Testing for serial correlation in least squares regression I. *Biometrika*, 37:409–428, 1950.

J. Durbin and G. S. Watson. Testing for serial correlation in least squares regression II. *Biometrika*, 38:159–178, 1951.

B. Efron and R. J. Tibshirani. *An Introduction to the Bootstrap*. Chapman and Hall, New York, 1993.

E. P. Ericksen, J. B. Kadane, and J. W. Tukey. Adjusting the 1990 Census of Population and Housing. *Journal of the American Statistical Association*, 84:927–944, 1989.

J. Fox. Effect displays for generalized linear models. In C. C. Clogg, editor, *Sociological Methodology 1987 (Volume 17)*, pages 347–361. American Sociological Association, Washington, DC, 1987.

J. Fox. *Regression Diagnostics: An Introduction*. Sage, Newbury Park, CA, 1991.

J. Fox. *An R and S-PLUS Companion to Applied Regression*. Sage, Thousand Oaks, CA, 2002.

J. Fox. The R Commander: A basic-statistics graphical user interface to R. *Journal of Statistical Software*, 14(9):1–42, 2005. URL http://www.jstatsoft.org/v14/i09.

J. Fox. Extending the R Commander by "plug-in" packages. *R News*, 7(3):46–52, 2007.

J. Fox. Aspects of the social organization and trajectory of the R project. *The R Journal*, 1 (2):5–13, 2009. URL http://journal.r-project.org/archive/2009-2/RJournal/2009-2/Fox.pdf.

J. Fox. *Applied Regression Analysis and Generalized Linear Models*. Sage, Thousand Oaks, CA, third edition, 2016.

J. Fox and M. Carvalho. The **RcmdrPlugin.survival** package: Extending the R Commander interface to survival analysis. *Journal of Statistical Software*, 49(1):1–32, 2012. URL http://www.jstatsoft.org/index.php/jss/article/view/v049i07.

J. Fox and M. Guyer. Public choice and cooperation in n-person prisoner's dilemma. *Journal of Conflict Resolution*, 22:469–481, 1978.

J. Fox and G. Monette. Generalized collinearity diagnostics. *Journal of the American Statistical Association*, 87:178–183, 1992.

J. Fox and C. Suschnigg. A note on gender and the prestige of occupations. *Canadian Journal of Sociology*, 14:353–360, 1989.

J. Fox and S. Weisberg. *An R Companion to Applied Regression*. Sage, Thousand Oaks, CA, second edition, 2011.

M. Friendly and P. Franklin. Interactive presentation in multitrial free recall. *Memory and Cognition*, 8:265–270, 1980.

A. Gelman, J. B. Carlin, H. S. Stern, D. B. Dunson, A. Vehtari, and D. B. Rubin. *Bayesian Data Analysis*. Chapman and Hall, Boca Raton, FL, third edition, 2013.

R. Ihaka and R. Gentleman. R: A language for data analysis and graphics. *Journal of Computational and Graphical Statistics*, 5:299–314, 1996.

T. Lumley and A. Miller. *leaps: regression subset selection*, 2009. URL http://CRAN.R-project.org/package=leaps. R package version 2.9.

D. S. Moore, W. I. Notz, and M. A. Fligner. *The Basic Practice of Statistics*. W. H. Freeman, New York, sixth edition, 2013.

J. A. Nelder. A reformulation of linear models. *Journal of the Royal Statistical Society. Series A (General)*, 140:48–77, 1977.

J. A. Nelder and R. W. M. Wedderburn. Generalized linear models. *Journal of the Royal Statistical Society. Series A (General)*, 135:370–384, 1972.

S. Papert. *Mindstorms: Children, Computers, and Powerful Ideas*. Basic Books, New York, 1980.

J. B. Ramsey. Tests for specification errors in classical linear least squares regression analysis. *Journal of the Royal Statistical Society Series B (Methodological)*, 31:350–371, 1969.

P. H. Rossi, R. A. Berk, and K. J. Lenihan. *Money, Work, and Crime: Some Experimental Results*. Academic Press, New York, 1980.

G. Snow. *TeachingDemos: Demonstrations for teaching and learning*, 2013. URL http://CRAN.R-project.org/package=TeachingDemos. R package version 2.9.

R. M. Stallman. *Free Software, Free Society: Selected Essays of Richard M. Stallman*. GNU Press, Boston, 2002.

H. A. Sturges. The choice of a class interval. *Journal of the American Statistical Association*, 21:65–66, 1926.

T. M. Therneau. *A package for survival analysis in S*, 2015. URL http://CRAN.R-project.org/package=survival. version 2.38.

T. M. Therneau and P. M. Grambsch. *Modeling Survival Data: Extending the Cox Model*. Springer, New York, 2000.

J. Tukey. Comparing individual means in the analysis of variance. *Biometrics*, 5:99–114, 1949.

W. N. Venables and B. D. Ripley. *Modern Applied Statistics with S*. Springer, New York, fourth edition, 2002.

S. Weisberg. *Applied Linear Regression*. Wiley, Hoboken, NJ, fourth edition, 2014.

G. N. Wilkinson and C. E. Rogers. Symbolic description of factorial models for analysis of variance. *Journal of the Royal Statistical Society. Series C (Applied Statistics)*, 22: 392–399, 1973.

Y. Xie. *Dynamic Documents with R and knitr*. Chapman and Hall, Boca Raton, FL, second edition, 2015.

Author Index

211

Subject Index

Printed in the United States
by Baker & Taylor Publisher Services

Printed in the United States
by Baker & Taylor Publisher Services